Stories in Stone

Garnet Books

Early Connecticut Silver, 1700–1840
by Peter Bohan and Philip Hammerslough
Introduction and Notes by Erin Eisenbarth

The Old Leather Man
by Daniel DeLuca

Westover School: Giving Girls a Place of Their Own
by Laurie Lisle

Henry Austin: In Every Variety of Architectural Style
by James F. O'Gorman

Making Freedom: The Extraordinary Life of Venture Smith
by Chandler B. Saint and George Krimsky

Welcome to Wesleyan: Campus Buildings
by Leslie Starr

Stories in Stone
How Geology Influenced Connecticut History and Culture
by Jelle Zeilinga de Boer

Stories in Stone

How Geology Influenced Connecticut History and Culture

Jelle Zeilinga de Boer

WESLEYAN UNIVERSITY PRESS
MIDDLETOWN, CONNECTICUT

Published by Wesleyan University Press, Middletown, CT 06459
www.wesleyan.edu/wespress

Printed in the United States of America
5 4 3 2 1

Wesleyan University Press is a member of the GreenPress Initiative. The paper used in this book meets their minimum requirement for recycled paper.

Library of Congress Cataloging-in-Publication Data
Zeilinga de Boer, Jelle.
Stories in stone / Jelle Zeilinga de Boer.
p. cm. — (Garnet books)
Includes bibliographical references and index.
ISBN 978-0-8195-6891-5 (cloth : alk. paper)
1. Geology—Connecticut. I. Title.
QE93.Z45 2009
557.46—dc22 2009011073

To Bjorn, Byrthe, and Babette,
with apologies for all those years
in which they had to share their father
with Connecticut's rocks

Contents

Preface ix
Acknowledgments xi
Introduction 1
1. In the Beginning: Continental Fusion and Breakup 8
2. Weather and Climate: Hurricanes and Ice Ages 27
3. Connecticut's Geologic Treasures: Gems and Ores 56
Sidebar: Gems in Quarry Tailings 74
Sidebar: Other Historic Quarries and Mines in Connecticut 82
4. Settlers and Soils in the Central Valley: The Legacy of Glacial Lake Hitchcock 83
5. The Metacomet Ridge: The Scientific, Political, and Cultural Impact of an Old Lava Flow 105
Sidebar: The Curse of the Black Dog 129
6. The Moodus Noises: The Science and Lore of Connecticut Earthquakes 132
Sidebar: Moodus Tremors and Sonic Booms 156
7. Visitors from Space: The Weston and Wethersfield Meteorites 157
Afterword: Our Lithic Inheritance 171
Bibliography 175
Index 197

Preface

Many of us have, at one time or another, wished that stones could speak. It first happened to me when I was six years old and was standing on a tropical beach holding a stone to my ear. I had just listened to a shell, and although it spoke in a whisper, the stone remained silent! My experiment was triggered by a volcano near my childhood home in Indonesia that had rumbled for weeks. I had been told that the rocks inside this giant had woken up and were plotting their escape. The piece of rock that I held in my hand was volcanic, and I was wondering what it could tell me about its birth and travels to the ocean. Only much later did I learn that the deep hum of an awakening volcano is caused by the rise and expansion of gas bubbles in magma. After the Japanese invasion during World War II and subsequent Indonesian revolution, I returned to the Netherlands with what remained of my family and decided to study earth science.

To find anything volcanic, Dutch geology students had to cross borders and travel to the Massif Central in France. There we found phonolites, a volcanic rock that emits a clear tone when struck. By varying the sizes of assorted fragments, we produced a xylophone and composed rock music. While working in Central America many years later, I learned about another, more sinister "voice" of the rocks: the rumble that precedes the shaking of an earthquake. Over time I learned that rocks could tell me so much, much more by their silence. Stones from space have provided us with information about the age of the solar system. The minerals in volcanic and metamorphic rocks testify to temperatures and pressures deep inside the Earth, and sediments speak about time, past climates, and evolution. In one form or another stones do speak, but only to those who are willing to listen. In this book I relate what rocks and their minerals have told me about the role they played in Connecticut history. They spoke to me, and I hope that they will do the same for you.

Acknowledgments

To friends, students, and acquaintances who contributed their time and energy to this book, I extend my deepest thanks. Ideas for *Stories in Stone* grew mainly out of observations made during field trips throughout the state of Connecticut. Several students became sufficiently interested to write term papers or theses, which greatly helped in gathering data and references. Most helpful were theses by Bob Altamura, Alison Guinness, Alyssa Lareau, Jan Ludwig, Carla Otfinoski, and Holly Shaw. Janet Stone tried to make me understand the intricacies of Connecticut's glacial history and also supplied essential figures. Tony Philpotts provided the photo of a thin section with Cobalt gold.

Steve Bischof and Suzy Taraba of Wesleyan's library were very helpful in locating old and, at times, rather obscure texts. Terry Prestasch drafted the line drawings, and John Wareham created the art work. John's patience is especially appreciated because too often I reorganized figures and requested changes. Gerrit Lekkerkerker, Lucie and Larry Iannotti, and Barbara Narendra struggled through early versions of the manuscript and helped to make it more readable.

Writing in a language other than my native Dutch was challenging and required much patience from my editors. Kathleen White succeeded admirably, and her work was further refined by Mary Crittendon. From early on, several years back, all chapters were hand-written and many pages resembled doctor's prescriptions. Felicité was somehow able to make sense of those notes and arranged them in logical order. She was indispensable.

To all I am very grateful.

Stories in Stone

Introduction

Connecticut—unique, peculiar, unmistakable
As in the smell of a skunk, the taste of an onion
Or the cry of a blue-jay screaming up the wind
—Odell Shepard, 1939

A person can learn much about Connecticut's landscape and past land use by taking a walk in the forests that cover its highlands. Follow any blazed trail and it becomes obvious that, with the exception of the northwest region of the state and the Central Valley, steep slopes and flat land are equally rare here.

Instead, gentle sloping hills, rolling like ocean waves and changing color with the seasons, predominate. Along the path a profusion of rocks and boulders stick out of the ground, as if they had grown like mushrooms. Usually a faint mineral layering is perceptible on their surfaces not covered by lichens. Bands of light and dark crystals alternate. Their attitude varies greatly from stone to stone, suggesting that the rocks once belonged to a coherent mass that was broken up, its fragments scattered across the land by surging sheets of ice.

Tall oaks and a few beech trees occupy the spaces between the boulders; somehow their roots have found sufficient soil in which to grow. These trees are all young, with straight trunks that rarely exceed a foot in diameter. Old gnarly trees with their wide canopies and massive branches, winners of past struggles for sunlight, are missing. The reason for their absence becomes apparent when the path cuts through a gap in a stone wall, built from glacial boulders, that crosses the forest, climbs a hillside, and stretches on to the horizon (plate 1). Not so long ago these low walls bordered open fields and meadows; by the end of the eighteenth century, in fact, more than three-fourths of Connecticut had been cleared and turned into farm and pasture land. New England households needed more than thirty cords of wood a year for domestic use alone, which represented the annual logging of an acre of mature forest. In 1834, in fact, Jedidiah Morse, in the account of his travels, wrote, "The whole state resembles a well cultivated garden." This rather romantic view, however, ignored the many negative impacts of wholesale deforestation and denudation, which caused changes in weather patterns and especially summertime droughts. Soils rapidly lost their natural nutrients, which were washed downhill, where they silted streams and rivers. Water tables fell, and the ecology changed irreversibly. Many plants and animals also disappeared, although, as the poet Odell Shepard writes above, skunks and bluejays remained. Hardly a cultivated Garden of Eden!

Significant regeneration, thankfully, has occurred in the last century. Forests have regained more then half of their original footing, but new land-use patterns continue to cause major problems. Houses, roads, and parking lots cover the best soils in the valleys and progressively crawl up the hillsides. As a result, clean freshwater, the state's most important domestic resource, has become increasingly depleted or contaminated. More then one and a quarter *billion* gallons of water are withdrawn daily; though much of it will return to the land, it will be contaminated when it does. Because of excessive pumping, some streams and small rivers carry barely any water at all during dry summers. So far, Connecticut has actually been lucky; rain has continued to fall, and the Connecticut River slakes our thirst with waters that accumulate first in Vermont and New Hampshire. However, past droughts show that this could all change rapidly. We can and should learn from the interaction between people and the land in the last four centuries in order to avoid past mistakes.

For early settlers, the seventeenth century was a period of much adaptation. They had to learn to deal with a different land and ecology. The soils in "old" England's agricultural south, where most of these people originated, contain a lot of lime, derived from soft bedrock that had formed in warm, shallow seas. Such soils are chemically basic and provide much-needed nutrients for raising grain. Soils throughout glacially ravaged New England, on the other hand, derive from hard quartz–rich formations; they are acidic and frequently very rocky.

Colonists had no experience raising native crops and early on decided instead to import cereals and grasses and experiment with those.

Without the annual addition of copious amounts of lime and fertilizers, however, "New" England's soils were not kind to European crops. The necessary agricultural adaptation would demand a great deal of time and energy.

Despite these difficulties, Connecticut became an agricultural powerhouse within a century. This was primarily due to the peculiar geology of the Central Valley, which stretches from New Haven to Deerfield, Massachusetts. It is drained by the Connecticut River and encompasses an area of about 1,300 square miles. Here, clays and silts had settled on the bottoms of large glacial lakes and covered the rocky debris left by sheets of ice. Eventually, the lake's natural dams broke, exposing their bottomlands. Those flat lands and fine lake sediments provided well-drained, relatively fertile soils that were in fact suitable to European farming practices when treated the right way. By 1775, then, Connecticut had become New England's breadbasket, exporting its agrarian products to rapidly growing cities along the Eastern Seaboard and to expanding plantations in the Caribbean.

The earliest mines and commercial quarries in the nation were developed in Connecticut (fig. I-1). These geologic resources they provided facilitated Connecticut's growth from a few colonies into a vibrant state. The glaciers that scoured New England had exposed extensive stretches of bedrock. After the ice melted, new drainage systems developed, resulting in many falls and rapids that provided a significant source of energy for mills. Large iron ore deposits in Salisbury, copper ores in Granby and Bristol, brownstone in Portland, feldspar in Glastonbury, and granite along much of the coastline provided Connecticut with resources that, when combined with water-generated energy and the proximity of major markets, propelled the young state into its industrial age (fig. I-2).

By the late eighteenth and early nineteenth centuries, Connecticut's industrial capacity had become a national phenomenon. Salisbury iron supplied guns and cannons for the Revolutionary and Civil Wars; Bristol's copper allowed for expansion of the famous clock industries of Thomaston and Plymouth; and Portland's brownstone provided the foundations for churches, mansions, and rowhouses in cities up and down the Eastern Seaboard. Stoneworkers carved granite for numerous gravestones and monuments, among them the Statue of Liberty, which stands on a pedestal of Connecticut granite.

Fig. I-1. The "Old Hole" in the southern section of Dodd's Granite Quarry, 1925. *Courtesy Willoughby Wallace Memorial Library, Stony Creek.*

The state's combination of resources, energy, and Yankee ingenuity gave rise to an amazing number of inventers and entrepreneurs as well. Best known are Samuel Colt (guns), Charles Goodyear (tires), Seth Thomas (clocks), and Eli Whitney (machinery). The variety of products manufactured in Connecticut was staggering, ranging from cotton textiles to rubber tires, knives to guns, sewing machines to washing machines, and carriages to automobiles.

In the twentieth century, sadly, Connecticut lost its agrarian independence and industrial eminence. Farms and pastures disappeared, to be replaced by asphalt ribbons and suburban agglomerations. Erosion rates rapidly outpaced natural soil formation. The mills and factories

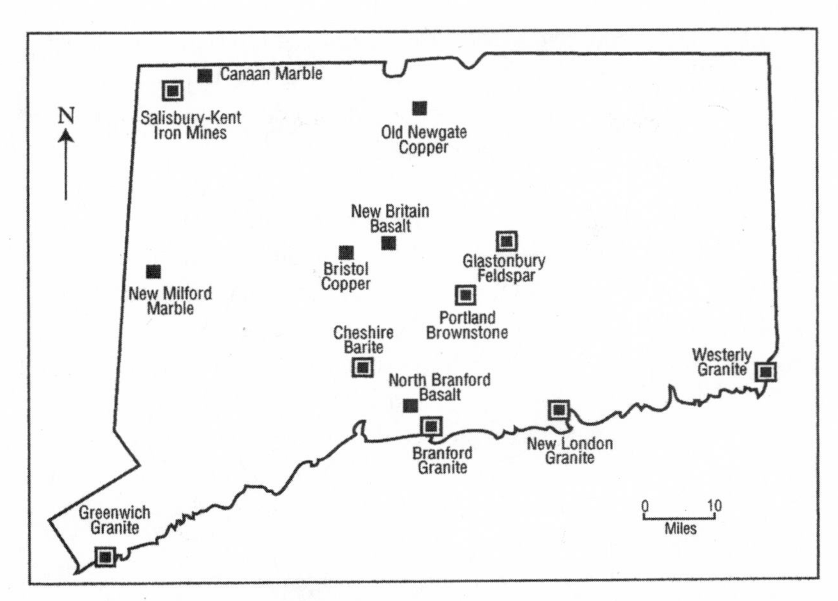

Fig. I-2. Historic mining and quarry operations in Connecticut. The Millstone granite quarry in Niantic started in 1648, the Portland brownstone quarry in 1665, and the Old Newgate copper mine in 1705. They were among the first commercial enterprises of their kind in colonial America.

built along the shrinking streams that once powered their engines crumbled into ruins. Hammers and steam machines fell silent. Of the more than seven hundred quarries that initially provided most of the state's geologic resources, only about twenty-five remain active. The Connecticut River, once western New England's most vital commercial artery, became a recreational "highway," clogged with pleasure boats on the weekends. Joe Cone, a native of Old Saybrook, wrote a poem about the river's trials more than a century ago:

> Still it sings the same sweet song
> And still it tells its tale
> Complaining of commercial wrong
> To forest, hill and dale
> It longs for freedom from the mills
> To be forever free
> To sweep unharnessed through the hills
> From cataract to sea.
> —Cone, 1901

The people of Connecticut clearly exhausted the region's resources early on. These days, almost four hundred years after the first settlers arrived, the state has "grown old" and relies increasingly on resources imported from other states and even foreign nations.

Volcanic, seismic, and climatic events have directly and indirectly played an equally important role in Connecticut's geologic impact on its human history. The remains of a 200-million-year-old lava flow provide the Central Valley with a green belt that stretches from New Haven to Amherst, Massachusetts. Its fractured trap rock collects rain, providing cities with clean water and stone for roadbeds and foundations. The ridge formed by this lava flow played a further role in the historic animosity between Connecticut's major cities, eventually leading the citizens of New Haven to fund the expensive and ill-fated Farmington Canal in the early nineteenth century.

Connecticut's "homegrown" earthquakes provided preachers with a means to control their flocks and scientists with a subject to argue about. Hypotheses on the quakes' origins range from growing carbuncles to drifting continents. The temblors remained quietly underground when the Yankee Atomic Power Plant was constructed above their realm in the Moodus area, but they returned with a vengeance a few years later, imparting an ominous warning.

Although hurricanes are generally considered a southern phenomenon, several have tracked north and penetrated the coastline where southern New England juts east into the sea. One of the worst, the New England Express, blew in on October 21, 1938, and caused extensive damage. Hundreds of homes were flattened and washed away along the shore; inland, buildings lost their roofs and church spires were shattered.

Among the most unusual of the state's geologic phenomena were the meteorites that invaded two Wethersfield homes a little more than a decade apart, in 1971 and 1982, preceded almost two centuries before by stones that rained from the sky in 1807 in southwestern Connecticut. Descriptions of the miracle by professors from Yale College led President Jefferson to state that he would rather assume the two Yankees were lying than accept that stones could fall from heaven.

This book begins with a description of Connecticut's geologic setting and climate, as seen through the eyes of its early settlers, its colonists, various Yankees, and a geologist. These chapters are followed by topical stories on the histories of rare gems/ores, glacial soils, lava flows, earthquakes, and meteorites. The hope is that *Stories in Stone* will help readers become more aware of the fascinating roles that natural resources and geologic phenomena have played in Connecticut's long history.

In the Beginning

Continental Fusion and Breakup

No one can understand Connecticut
Who leaves the rocks out of his reckoning,
Three hundred years, now, we have worked among them,
And they have worked on us to more effect—
—Odell Shepard, 1939

Rocks were and continue to be essential to life on Earth. When molten rock escapes from fissures and volcanoes, it carries oxygen and hydrogen, which combine as water vapor. These steamy clouds condensed and filled lakes and oceans over billions of years, painting the planet blue. Once at the surface, the igneous rocks, their volcanic offspring (the volcanic rocks that form when the magma reaches the surface), and their metamorphosed neighbors (older rocks around the throats of the volcano that are heated) disintegrate. The elements inside the rocks' minerals escape their crystal prisons to become part of the soils, providing essential nutrients for plant life and in turn supporting the animal kingdom.

Writers once in a while recognize the virtues of rocks. In Hermann Hesse's famous novel, the title character, Siddhartha, says, "This is a stone, and within a certain length of time it will perhaps be soil and from the soil it will become a plant, animal or man." Annie Dillard writes, "Rocks shape life, and then life shapes life . . . life and the rocks, like spirit and matter, are a fringed matrix." However, most of us continue to view rocks as a boring nuisance. Farmers curse the rocks in their fields; highway builders blast them out of their chosen paths. Few people realize that without rocks and the materials derived from them, modern industry itself would come to an abrupt halt. Just about everything we use comes directly or indirectly from under the ground. Whereas cave dwellers needed just a few pounds of stone to fashion their spear points and arrowheads, contemporary Americans "consume," on average, more than twenty thousand pounds of stone, sand, and gravel per person annually and more than one thousand pounds of iron and steel. Rocks are literally the "bedrock" of our civilized world. As twentieth-century historian Will Durant points out, "Civilization exists by geological consent."

When compared to the topography of northeastern states such as New York or New Hampshire, Connecticut's monotonous hills appear uninspiring (plate 2, top). Connecticut lacks the Finger Lakes, the Adirondacks, or the White Mountains. Because topography can reflect underlying geology, we might expect equally little diversity in the state's rock types. The opposite is true, however: more rock types of different composition, texture, and origin are exposed in relatively small Connecticut than in most other states in the nation. Outcrops of bedrock commonly appear on ridge crests, in stream valleys, and along highways as reddish or grayish masses, implying a boring uniformity. On closer examination, though, they reveal almost painterly qualities—the boldly spattered stripes of Pollock, the smooth curves of Rubens. The stripes represent sediment layers deposited many millions of years ago in the shallow waters of bygone tropical seas; the curves are reminders of the crushing collisions of huge continental plates, which heaved and folded those sediments into today's mountain ranges.

Geologic Time

> Some drill and bore
> The solid earth and from the strata there
> Exact a register by which we learn
> That he who made it and revealed its date
> To Moses, was mistaken in its age.
> —William Cowper (1731–1800)

For many centuries, enduring religious beliefs guided ideas about the age and origin of the rock formations at the Earth's surface. According to the Book of Genesis, God "constructed" Earth and everything on it in six days, and then rested on the seventh. Theophilus of Antioch was probably the first to use Genesis to try to estimate the age of the Earth; working around the year 180 CE, he concluded that creation began in 5515 BCE. In the seventeenth century James Ussher, an Irish bishop, used an elaborate genealogical study of individuals mentioned in the Bible to conclude that the Earth was instead created in the year 4004 BCE, and one of Ussher's followers, John Lightfoot, went so far as to propose the exact date and time: October 23, at 9:00 a.m.! In Ussher's scheme, Adam and Eve had a little over two weeks to enjoy Paradise before their expulsion on Monday, November 10. On May 5, 2348 BCE, the ark ran aground on Mount Ararat, and the Earth was repopulated.

Ussher's theory had a catalyzing effect on European geologists, who now began to recognize the concept of extended time in various rock formations. The most famous among them was James Hutton, an eighteenth-century Scottish naturalist who understood the cyclic nature of geologic processes and envisioned the ruins of earlier worlds in the folded strata of the hills. His field studies showed that erosion ultimately wears down mountains, and the soft sediments produced by this process accumulate in low-lying areas, where they are in turn slowly buried, layer by layer. Hutton theorized that this burial process would result in the heating and recrystallization of minerals, which in turn produce rocks. Lastly, these metamorphosed masses were thrust from their depths back up to the surface, resulting in new mountains. Hutton concluded in his *Theory of the Earth* (1795) that such cycles had occurred repeatedly and showed "No vestige of a beginning, no prospect of an end."

But there had to have been a beginning, and the search was on! Hutton's French contemporary, the Count de Buffon, experimented with a series of heated iron spheres of different diameters, assuming Earth to have been a molten globe at its origin. After carefully timing and comparing the cooling rates of iron spheres with various diameters, he extrapolated an age of 74,832 years for the Earth, more than twelve times Ussher's estimate. Lord Kelvin, the famous British physicist, considered Buffon's approach childish. Kelvin, whose method was based on more advanced scientific principles, also calculated the time for the Earth to cool, first coming up with an age of about ninety-five million years. He kept revising his numbers, however, and in his last paper on the subject, in 1897, he reduced his estimate to about twenty-four million years. Despite plenty of controversy, then, the estimates of the Earth's age had begun their inevitable increase from a few thousand to many millions of years.

The earth sciences entered the American academic world in 1801, when Timothy Dwight, the president of Yale University, offered Benjamin Silliman the newly created Chair of Chemistry and Natural History. Of his appointment he wrote, ". . . it was the cause of wonder to all and of cavil to political enemies of the college." Silliman, originally educated as a lawyer, trained for his new career with chemists in Philadelphia and geologists in England. He published his first paper, "Sketch of the Mineralogy of the Town of New Haven," in 1810. Silliman believed that the Hebrew word *yum* in Genesis referred to a millennium rather than a day. Creation would then have taken place over six thousand years, while the seventh millennium (of rest) was still unfolding. He described the geological formations around New Haven as the product of "the long progress of 6,000 years," concluding that "for many ages, if not from creation, things have remained substantially as they now are."

The statement appears to have been written mainly to appease Ussher's followers on the Yale faculty. In a paper published twenty-three years later and titled "Consistency of the Discoveries of Modern Geology with the Sacred History of the Creation and Deluge," Silliman clearly supported Hutton's ideas and wrote, "the days of Creation were periods of time of indefinite length." Most Yankees, however, continued to believe in Ussher's time frame. In 1835, a Mr. Draper, stumbling upon dinosaur tracks exposed in brownstone slabs used for a walkway in Greenfield, Massachusetts, exclaimed, "Here are some turkey tracks made 3,000 years ago."

Outcrops in the New Haven area show small fragments of schists, or metamorphosed clays, from the Western Highlands inside layers of brownstone. Younger volcanic masses in turn intruded up on and then covered the sediments. Such events provide relative ages for the different rock formations—the metamorphic rock had to be older than the sediments, and the volcanic rock, younger. Using this comparative methodology, geologists constructed a basic chronology of many of Connecticut's geologic formations by the end of the nineteenth century. Without absolute ages, however, geologists could not anchor their chronology to the four specific geological periods of the European paleontologists, who studied the fossil content of sedimentary rock formations. These were the Cryptozoic ("hidden life"), the Paleozoic ("old life"), the Mesozoic ("middle life"), and the Cenozoic ("new life") (see fig. 1–1). Paleozoic rocks dominate in the Eastern and Western Highlands, while Mesozoic sediments and volcanic rocks underlie the Central Valley, which bisects Connecticut from north to south (plate 2, top). Although those rock formations were easily attributed to these broad geologic eras, their absolute ages remained unknown.

In 1889, geologists working for the United States Geological Survey found uraninite crystals in igneous rocks quarried near Glastonbury. Seven years later Henri Becquerel, a French physicist, discovered the process of radioactive decay. Expanding on Becquerel's work, Ernest Rutherford, a nuclear physicist then at Cambridge, hypothesized that during the radioactive decay of uranium, helium gas became trapped inside the crystals, slowly leaking out over time. Perhaps the amount of gas remaining in a given sample could provide an "absolute" age for it. Bertram Boltwood, a radiochemist at Yale, designed the laboratory instruments to make the necessary measurements that validated Rutherford's hypothesis. In a November 18, 1905, letter to Rutherford, Boltwood reported his first results for a collection of uranium-containing rocks. About half the samples Boltwood used had been collected from Connecticut pegmatites,[1] rocks of the granite family. Five rocks from outcrops near Portland and Glastonbury supplied an average age of 95 million years, while three from Branchville (in western Connecticut) were estimated to be about 125 million years old.[2] Suddenly it became obvious that the ages of certain rock units

1. Pegmatite is a light-colored, igneous rock, composed mainly of relatively large crystals of feldspar and quartz and containing many rare minerals.

2. After additional corrections, Boltwood concluded in a paper he wrote two years later that the rocks were actually much older, 410 and 535 million years, respectively. Shortly thereafter it was discovered that helium, because of its small molecular size, escapes its containment at a fairly rapid and uneven rate, so this dating method is not reliable.

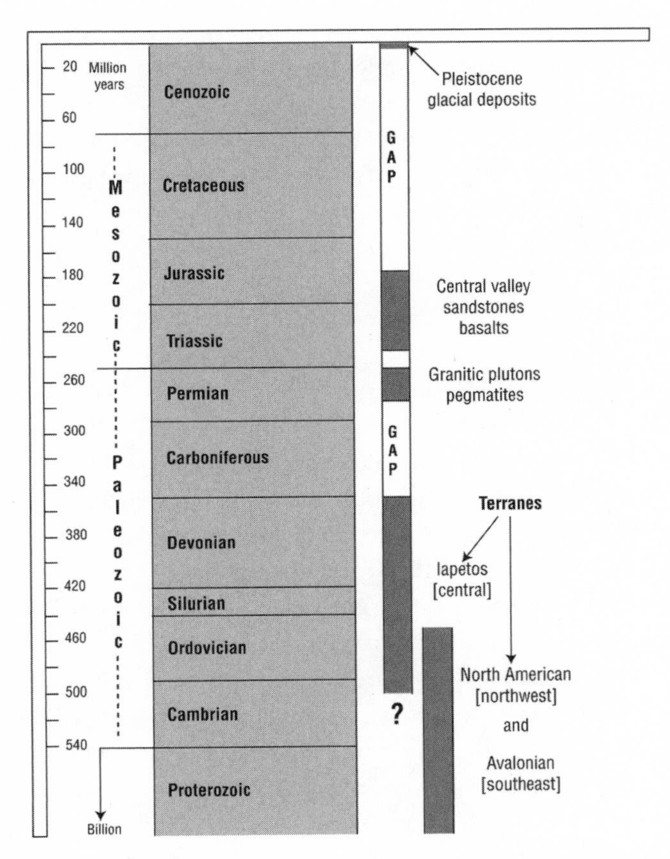

Fig. 1–1. Simplified stratigraphic column with principal geologic eras/periods and approximate age span of historic geologic terranes/entities in Connecticut.

should be measured not in thousands but in millions of years, just as Lord Kelvin had predicted. Those time frames jolted the scientific community.

In the following decades, new dating methodologies were developed and refined. Further work on the Glastonbury pegmatites showed them to be in fact about 255 million years old. When geologists began to date samples collected in the interior of the North American continent, it became obvious that some rock units were billions of years old. Visitors from outer space ultimately provided the true age of the Earth. Meteorites average around 4.56 billion years old, which is believed to be the time in which the entire solar system came into existence.

Millions, let alone billions, of years are extremely difficult to comprehend, so the overall span of geologic time has often been likened to the passage of a single year. In such a scheme, the age of Connecticut's oldest rocks is barely ten weeks, and humans themselves have only witnessed a small part of the final hour of December 31. Yet in these last few seconds before midnight, they have radically transformed Earth's surface and even accelerated some of its geologic processes.

Geologic Mapping

Many eons ago, Connecticut's mountains were at a scale that rivaled that of the Alps. Their raw-boned peaks pierced the clouds that shrouded their tree line. The Appalachians are old, older than present-day ocean floors and most other mountain chains on Earth. Judging from the depth from which their roots eventually rose, they may have reached their highest levels in Connecticut. All that is left are highly metamorphosed rocks, and it is from these remains that geologists have to estimate how long it took for the mountains to grow and fade back into hills.

The oldest rock formations in Connecticut, exposed in the northwestern corner of the state, are more than one billion years old. The reason for Connecticut's relatively young geologic age rests in the cyclic nature of geologic events and the region's former location at the edge of the North American tectonic plate. After soils and sediments erode from mountain ranges, they accumulate in offshore basins along the margins of continents, where they are buried deeply and slowly transformed into rock. When continents collide, these masses are forced up, fractured, and folded, forming new mountain ranges, such as the Appalachians. As soon as this uplift occurs, erosion takes over, and the resulting sediments are moved to basins that form along the new margin of the continent. In several such basins, including those along North America's east coast, as much as five miles of sediment have accumulated in the last 200 million years. There they will remain until the plates collide again, many millions of years from now.

Clarence King, a Yale graduate and the first director of the United States Geological Survey, was among the early scholars who described the "pulse" of the Earth. In 1877 he wrote, "The ages have had their [long] periods of geological serenity, when change progressed in the still, unnoticeable way, and life through vast lapses of time followed the stately flow of years, drifting on by insensible gradations through higher and higher forms, and then all at once a part of the earth suffered a [relatively] short, sharp, destructive revolution." Such revolutions are known as orogenies, or periods of mountain building.

Three orogenies shaped Connecticut during the Paleozoic period. The first mountains to appear were the Taconics of eastern New York state and westernmost Connecticut. The next phase of mountain building was the Acadian orogeny, which affected all of Connecticut. The concluding Alleghenian orogeny provided geologic formations and structures concentrated mainly in eastern Connecticut and Rhode Island. Thus, over a period of about 250 million years, three partially overlapping strips of continental crust were added to the much older basement of North America, in much the same way growth rings annually widen the girth of a tree trunk. All continents grow in this manner. During the Paleozoic period, North America grew mainly on its eastern side; during the Mesozoic and Cenozoic periods, it grew to the west.

Where tectonic plates collide or separate, volcanism occurs. Voluminous granitic masses and numerous volcanic veins attest to widespread magmatic activity in Connecticut during the Paleozoic orogenies. Thick basaltic lava flows in the Central Valley testify to the embryonic phase in the opening of the Atlantic basin during the early Mesozoic period, when the North American and African plates broke apart.

The general interest in Connecticut geology began as soon as settlers discovered deposits of useful metals such as copper, iron, and lead in the late seventeenth century. True geologic mapping, however, did not start until the first half of the nineteenth century. Courses taught by Benjamin Silliman at Yale, and his founding of the *American Journal of Science* in 1818, aroused great interest in the study of natural history in America, which led to the institution of geologic instruction in other colleges and the establishment of state geological surveys in several eastern states. The first geological survey of Connecticut began in 1835, when Governor Henry Edwards contracted with Charles Shepard and James Percival to undertake it. Shepard was to concentrate on mineralogy and economic geology, while Percival was to identify and map the rock formations.

The two men traveled on foot and horseback, ultimately surveying the entire state from east to west at two-mile intervals, which was a slow, arduous endeavor. Shepard's task was easier because, by 1835, many mineral and ore locations were known and some actively worked. He dutifully submitted his report after only six months on the job.

Percival's task was much more difficult and took considerably longer, despite the fact that by this time deforestation had exposed large areas of rock. Once he had located a specific and recognizable formation, he had to track it over hills and through valleys to connect exposures that were often separated by long stretches where glacial sediments hid the rock. In 1841 state legislators, tired of waiting for Percival's results, refused further funding. Percival (fig. 1–2) then finished his "Report on the Geology of Connecticut" in 1842. On his map, he delineates most major rock types with an accuracy that leaves even contemporary geologists speechless. These days, geologists might spend several years mapping a single quadrangle. Percival surveyed all ninety-nine in just seven years.[3]

The second state geologic map was published in 1907 by Herbert Gregory and Henry Robinson, and the third in 1985 by John Rodgers (plate 2, bottom).

Geologists are taught to read outcrops like paragraphs in a book; unfortunately, a number of pages are invariably missing, appearing later in other outcrops, as if the winds had blown them across the land. It takes imagination to put those pages back in the correct order, and the older they are, the more difficult the task, because those pages are the most crumpled and worn. Already in 1876, Clarence King had concluded, after extensive fieldwork, that geology itself is "chiefly a matter of the *imagination*—one man can actually see into the ground [only] as far as another." Even with the most modern and advanced exploration tools, King's dictum remains true. Geology to a large extent resembles art.

One of the most difficult aspects of geology is developing an accurate stratigraphy, or sequence of deposition and/or emplacement of rock formations over time. In Connecticut's Central Valley, this task is relatively easy, because the valley's contents dip east, therefore exposing the oldest

3. After having received his degree in medicine in 1820 from Yale, Percival spent a few years as surgeon at West Point. He subsequently became a prolific poet, newspaper editor, and proofreader/assistant to Noah Webster.

Fig. 1–2. James Gates Percival. Eminent poet, accomplished linguist, and outstanding geologist.

formations along its western margin. The metamorphic[4] rocks in the Highlands, on the other hand, were severely contorted by several tectonic events. There, tracing the convoluted surface patterns of specific formations relies on the presence of specific characteristics, such as color, texture, and the occurrence of certain "guide" minerals. Because of repeated folding and fracturing, this is a difficult task (fig. 1–3), and most geologic maps are therefore hypothetical reconstructions that are often influenced by geologic "fashion."

4. Metamorphism is the transformation of soft sediments into rocks due to increases in temperature and pressure during burial. For example, clays become slates or schists, and limestones change into marbles.

Fig. 1–3. Examples of folding by compression and faulting by crustal extension.
Top: Conjugate set of steeply dipping faults enclosing a small rift zone (Route 11, exit 4, near Salem).
Bottom: Asymmetric flow fold, showing the ductility at deep levels in the crust during its formation (Route 9, exit 9, near Higganum).

Geologic History and Key Outcrops

Unraveling the sequence of deposition, metamorphism, and deformation of rock units is difficult because single outcrops usually provide information only on a thin slice of time. Fortunately, "key" outcrops allow for the differentiation of several separate geologic events in a single exposure.

For example, the aforementioned James Hutton, one of the founders of modern geology, located two outcrops of singular importance in Scotland that helped put to rest a major controversy among "Plutonists" and "Neptunists" in the late eighteenth and early nineteenth centuries. The Neptunists believed that all rocks had originated below sea level as mechanical and chemical deposits. The Plutonists asserted that part of the Earth's crust was of igneous (volcanic) origin and was related to the surfacing of deep-seated molten masses. The first key outcrop Hutton found in this regard was in the Siccar Cliff, east of Edinburgh. There, horizontal beds of red sandstone overlie a vertically layered formation of grey mica schists. It was clear to Hutton that a long interval, called a

Plate 1. Old stone wall in a reborn forest, Deep River area.

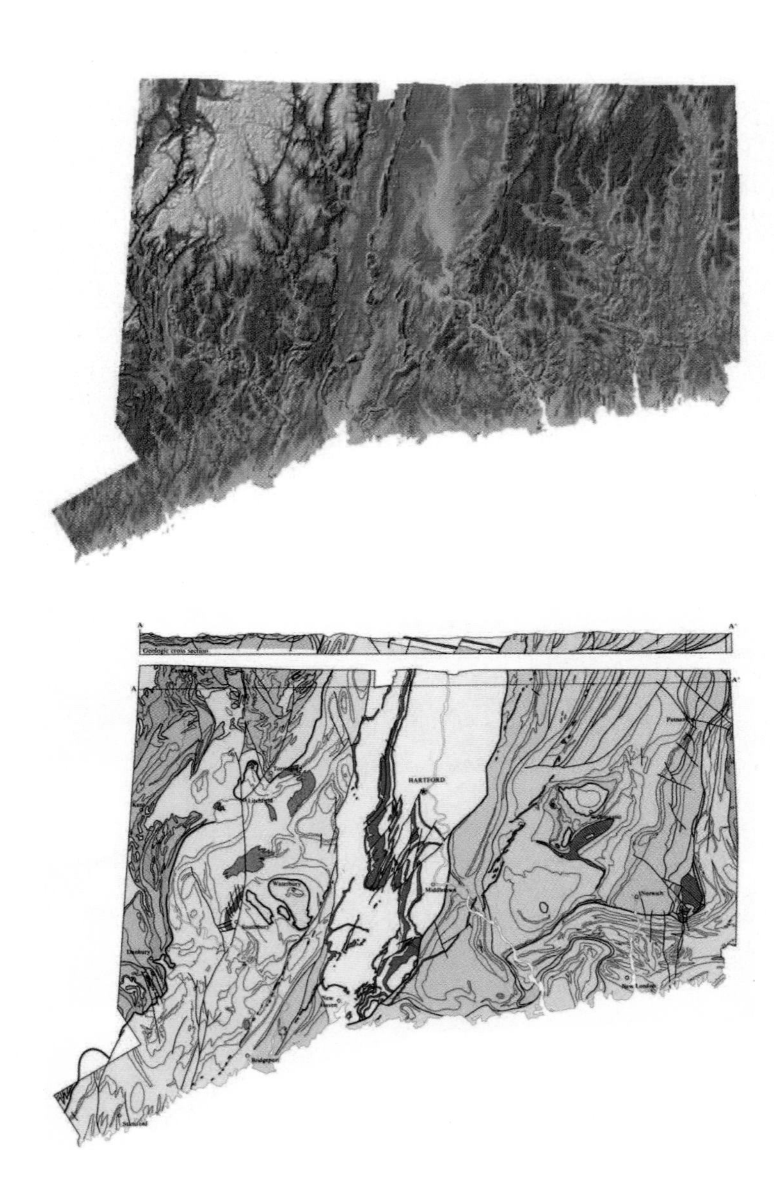

Plate 2. Top: Topographic relief map of Connecticut. *Courtesy Chester Arnold; Center for Land Use, Education and Research, University of Connecticut.* Bottom: Geologic Map of Connecticut. *The Connecticut Geological and Natural History Survey, 1989.*

Plate 3. Detail of the Great Unconformity in the Roaring Brook of Southington. Steeply inclined Ordovician schist is overlain by gently east-dipping sandstone layers of Triassic age.

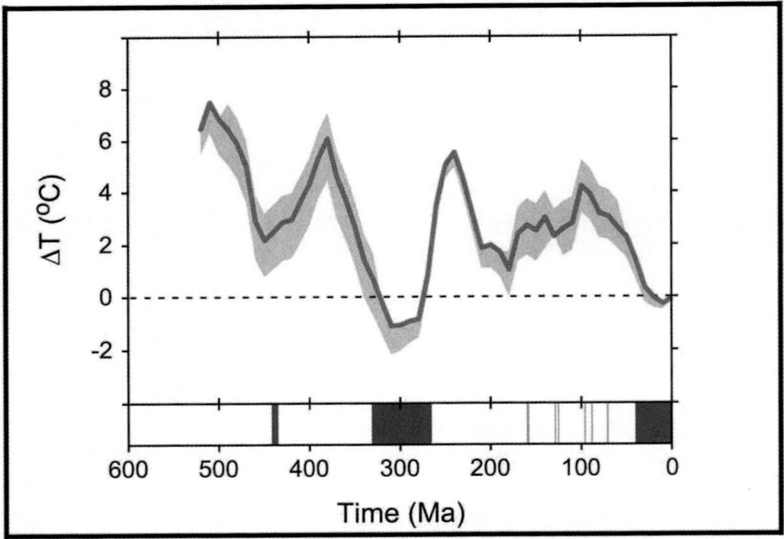

Plate 4. Top: Exposure of early Jurassic sandstones along Route 9, exit 22, in Berlin, showing the black lake deposits sandwiched between red beds that were deposited during dryer climate conditions in the same subtropical environment. Bottom: Variations in global climate in the last 500 million years. The wide blue bars indicate glacial periods, the narrow ones cold periods. *Courtesy Dana Royer, Wesleyan University.*

geologic discordance, separated the two rock units. The schists, which had originated as layers of clay on a seafloor, had been buried, metamorphosed, and folded before being forced back to the surface. Erosion above sea level had shaved the upper sections of the tilted schists, and red layers of sand, derived from nearby mountains, had been deposited on this surface much later. Clearly, there had not been continuous deposition below sea level, as the Neptunists proposed.

Hutton's second key outcrop is located in the gorge of the Tilt, a stream that deeply intersects the Grampian Mountains. At this site, steeply dipping granite veins slice through a thick section of mica schists. To Hutton, those veins were proof that the granite had an igneous, not an aqueous, origin, and that it had risen from below. The veins provided evidence for the high temperatures at depth that are necessary to transform clays into mica schists. Hutton became so excited at this proof of his hypothesis that the guides who accompanied him and his friend John Playfair, who described this scene, were convinced that he had discovered gold, for that alone seemed likely to explain his "strong marks of joy and exultation."

Connecticut has a geologic key of similar importance in the Roaring Brook of Southington. A single outcrop in its steep ravine resembles almost exactly the two exposures that so impressed Hutton. At this site, a layer of Mesozoic sandstone covers a steeply inclined sequence of Paleozoic mica schists, intruded by pegmatite and quartz veins. The contact between these formations is undulatory, and it is known as the "Great Unconformity" because evidence for a significant segment of geologic time, as much as 200 million years, is missing between the vertical schists and the horizontal sandstone layers (plate 3).

In the following I will discuss the great wealth of geologic information that one can garner from a single key exposure, especially when it is combined with knowledge about New England's plate-tectonic history that has been obtained in the last four to five decades. The story covers a time frame that stretches from approximately 440 million years ago to the present, and from clays washed into an old ocean, long gone, to clays splashed downstream during construction of modern houses on the hill site.

• In the late Ordovician and Silurian periods (fig. 1–1), the erosion of the Taconic Mountains created sediments, thick layers of silt and clay, that were deposited in deep basins along the margin of the North American plate. The Taconics had resulted from an earlier plate-tectonic collision sometime between 480 and 440 million years ago, and the Hudson Highlands, Berkshire Hills, and Green Mountains now represent the roots of that ancient, deeply eroded range.

• The Iapetus Ocean,[5] which separated the Laurasian (North America and Eurasia) and Gondwana (Africa and South America) continents gradually closed during the Devonian period. Its seafloor was thrust below the North American continent, and absorbed into the Earth's mantle. When the continental segments of the plates crashed into each other sometime in the Middle Devonian, the sediments that had accumulated in offshore basins were squeezed upward, and a new mountain range, the Appalachians, rose. At depth, new minerals had grown in those Late Ordovician/Silurian sediments, transforming the clays and silts into mica schists containing silvery muscovite (mica) flakes and reddish-brown garnet nodules. Layers of the schists at different outcrops in the Roaring Brook ravine tilt at various angles. Their attitudes reveal asymmetric folds with northerly trending axes that resemble petrified waves breaking on North America's ancient shore, indicating that the principal force responsible for the deformation came from the east.

• The buried sediments in offshore basins that had reached the deepest levels were at such high temperatures and under such high pressure that they started to "sweat." This process, known as partial melting, caused quartz and feldspar grains to disintegrate. Their basic elements combined with water, resulting in silica-enriched melts and hot fluids that accumulated and slowly rose into overlying rock formations. The subvertical layering of the Roaring Brook schists provided ready-made pathways along which those melts and fluids could force their way upward. Pegmatite veins dated at 345 ± 5 million years (the early Carboniferous period) formed after the melts cooled, and vertical quartz veins represent the injection of the somewhat younger fluids.

• After an unknown period of time, a steep fault that developed parallel to the foliation of the schists intersected with and offset one of the pegmatite veins. The fault first followed the eastern contact between

5. Iapetus was the father of Atlantis, hence the use of his name for the Paleozoic ocean that preceded the much younger Atlantic.

pegmatite and schist, then cut across the vein and followed the western contact. Horizontal slippage along this fault caused the eastern section to move northward relative to the western side, resulting in a local doubling of vein width. Fractures along which such horizontal motion occurs commonly develop in regions where plates slide along each other, as in present-day California. Similar faults occur in a broad north–south trending zone that stretches from West Haven into Vermont, suggesting a regional tectonic event. This California-type deformation most likely occurred around 300 million years ago, during the initial phase of the Alleghenian orogeny, the third and last of the three Paleozoic periods of mountain building in New England.

• In Permian time (290 to 250 million years ago) Connecticut was completely surrounded by land masses and located in the middle of the North American plate of a supercontinent, called Pangaea, which stretched from pole to pole. The Earth at that time had only one huge continental mass and a gigantic ocean, Panthalassa. This assembly of plates had a major effect on the chemistry of ocean waters and Earth's climate; more than 90 percent of all marine species and 70 percent of terrestrial vertebrates had become extinct by the end of the Permian.[6] Paleontologists used this relatively abrupt ecological change to establish the end of the Paleozoic and beginning of the Mesozoic era. No Permian sediments remain in Connecticut, but large granitic masses and associated pegmatite and quartz veins were emplaced. The largest granitic masses of this age, known as plutons, occur in a wide east–west trending zone that stretches from the Trumbull area in western Connecticut to the Narragansett region of Rhode Island. A series of plutons with similar age exists in Cornwall, southwest England, and may represent the eastward continuation of this belt before the opening of the Atlantic Ocean.

• About 20 million years after the end of the Permian period (some 230 million years ago) (fig. 1–1), Pangaea broke apart and the North American and African plates began to separate, a process that led to the birth of the Atlantic Ocean. Preexisting fractures were reactivated in the Appalachians, and vertical slippage along these faults led to the formation

6. This catastrophe was most likely due to severe climate changes caused by a combination of destructive events including global cooling, a possible meteorite impact near Antarctica, and voluminous volcanism in Siberia.

of rift valleys. Both east and west of the Roaring Brook outcrops, northerly trending faults outline the western margin of the Connecticut rift zone (the Central Valley, plate 2). Streams from the Western Highlands cascaded across these faults on their way to the lowlands, carrying much sediment. Because rock units at Roaring Brook contain pegmatite and quartz veins, which are more resistant to erosion than schists, an irregular, undulatory surface, the "Great Unconformity," developed (plate 3). Coarse sands with rounded pebbles accumulated on this erosional surface. About 180 million years ago, the region tilted east, suggesting the uplift of the Western Highlands. In the following millions of years, geologic time appears to stand still in New England. Erosion dominated the period as the Appalachian Mountains slowly changed into the hills that characterize Connecticut today.

• Glacial till, an unsorted mixture of material scraped from the bedrock farther north, blankets the hills on either side of the stream. Much of the till was deposited some nineteen thousand years ago, when the ice sheet that covered the region melted and the debris it had carried southward dropped out. After all of the ice had left, the Roaring Brook cut through the till and washed the glacial sands and clays downstream, leaving the large boulders that still litter its streambed. At the falls, the stream cut completely through the glacial till and underlying Mesozoic sandstone layers into the schists below, revealing the unconformity.

• Where the stream cut into its right bank, about fifty yards east of the falls, it eroded the sandstone, opening an overhang that exposed a steeply inclined, smoothly textured fault. Fracturing was probably associated with deformational processes during the late Jurassic period, when voluminous masses of basaltic magma forced their way up along the western border of the rift zone. The space below the overhang provided an excellent temporary shelter for native people, who used the quartz pebbles in the stream to produce arrowheads and other implements.

• About a decade ago, new houses were constructed on the hilly land southwest of the stream. Excavation disturbed the glacial till layer, and heavy rains washed much of its reddish clay and sand downhill into the stream valley. Those sediments covered the exposed schists just as the more than 200-million-year-old sandstones had done when the Great Unconformity was created. A few months later, heavy rain removed all traces of that event at the falls, but the new sediments washed downstream, covering older sand and clay layers in a gradually thickening package at the foot of the hill. Thus, as the French say, "L'histoire se répète."

This single outcrop in the Roaring Brook provides evidence for several major tectonic events that shaped Connecticut, including a compressive event caused by colliding plates, an oblique collision when plates brushed along each other, and a period of crustal spreading as the plates separated. This all happened in about 400 million years, less than 10 percent of the Earth's total age.

Connecticut's Tectonic Framework

Using a quite simplified model, we might see Connecticut's tectonic framework as resembling a house with a corrugated roof, a sunken living room, and a basement (fig. 1–4).

The roof consists of a thin layer of glacial and post-glacial sediments. These tills and lake sediments are draped over hills and valley bottoms and cover much of the state. On steep slopes and hill crests they have often been eroded, leaving bare bedrock or fields of boulders. Man-made walls made of those cobbles and boulders cover the glacial deposits in intricate patterns that appear to stitch the landscape together. In an architectonic sense, such cover can be called a leaky roof. Only the thicker lake deposits of the Central Valley are shown (fig. 1–4, top).

The basin that slashes through central Connecticut formed during the early Mesozoic period and represents its sunken living room (fig. 1–4, middle). Its floor sank as much as three miles over a period of about fifty million years. During its slow subsidence, this elongate depression filled with sands and clays washed in from the Eastern and Western Highlands. During their burial, the soft sediments consolidated into sandstone and shales. The Central Valley also contains the remains of three lava flows. The striking contrast between the black/orange basalt flows and the reddish-colored layers of sandstone is striking and exposes nature's decorating ability.

The foundation of Connecticut is indeed massive. Its basement complexes extend to a great depth, where the Earth's crust changes gradually into

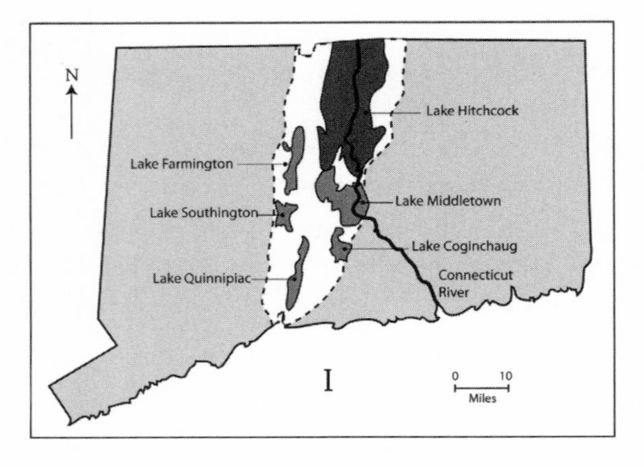

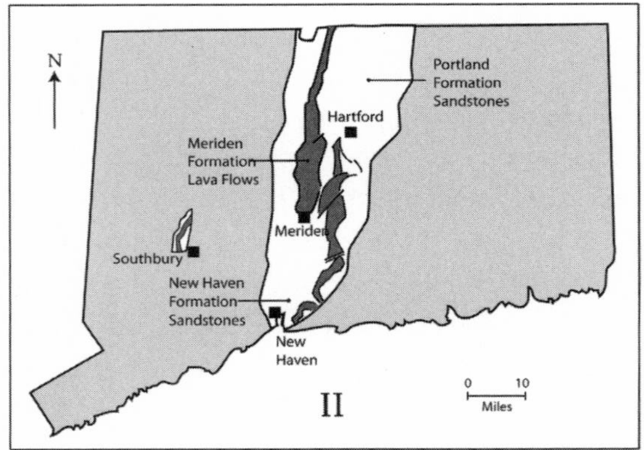

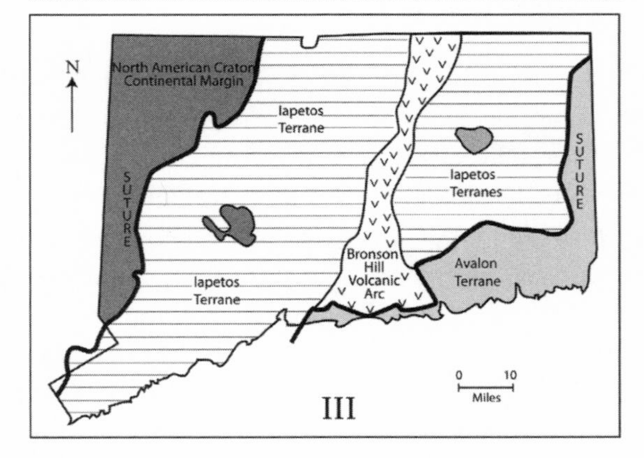

Fig. 1–4. Connecticut's geologic framework, much simplified. I: Pleistocene glacial lake deposits in the Central Valley; II: early Mesozoic Central Valley (rift zone); III: Paleozoic terranes and principle suture zones (basement).

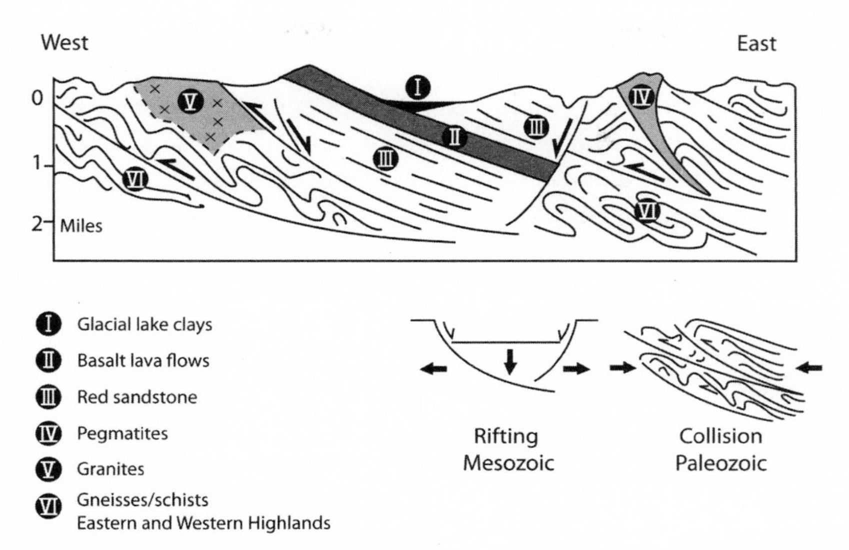

Fig. 1–5. Schematic cross-section (east to west) of central Connecticut with principal rock units and structures. Compression (folding) occurred at three different times in the Paleozoic, and crustal spreading dominated in the early Mesozoic.

its mantle.[7] The exposed formations that make up Connecticut's basement are divided into several geologic terranes, segments of Earth's crust with different origins and geologic histories that were glued together during the periods of mountain building.

Geological terranes in the northwest contain the metamorphosed remains of a shallow subtropical sea that washed the shores of the North American landmass during the early Paleozoic period. Most characteristic of the sediments that accumulated in this region are shallow-water limestone complexes that were changed into marbles. Similar sediments form today in the warm waters surrounding the Bahamas and Bermuda.

The terrane in the middle of the state consists mainly of layered rock formations (former sediments) that originated in the deeper waters of the Iapetus Ocean, seaward of the original continental slope. Clays and silts dominated these deposits, but they also contained airborne products from a nearby active volcanic arc. All were deeply buried, metamorphosed, and then lifted during the collisions of the continental plates. In the following

7. The mantle is the part of Earth between the crust and the core, and it comprises most of Earth's volume.

eons, erosion shaved layer after layer from the newly formed mountains, unloading the pile in a process that can be compared to the melting of an iceberg, whose deeper sections become visible as its surface shrinks away. In the past several hundred million years, rock units from the Appalachian root zone have surfaced and become exposed in the Highlands. Many show the effects of the high pressure and temperatures they experienced at depth, where they acquired the consistency of toothpaste before hardening again as they rose (fig. 1–3, bottom).

A very early immigrant known as the Avalon terrane occupies the southeast corner of Connecticut. Its rock formations are older and more highly contorted than those of the former Iapetus Ocean and appear to be predominantly of volcanic origin. Plate-tectonic reconstruction suggests that this crustal segment, which included a volcanic arc, might have been located in the southern hemisphere, along a margin of the African continent. It separated and slowly drifted north to dock with the North American continent. Avalon most likely consisted of a very long, slender sliver of crust similar to present-day California. Additional segments of Avalon have been found in the Carolinas, Rhode Island, eastern Massachusetts and eastern Maine. Avalonian crust also occurs in southern Ireland and Wales, suggesting that the sliver was a coherent entity before its northern and southern segments separated during the opening of the Atlantic basin in the Early Mesozoic period.

Figure 1–5 provides a schematic cross-section from east to west across central Connecticut. It shows the combined effects of several collision events (folds) in the Paleozoic period and the crustal extensions (faults) during the Mesozoic period. Two wide fault zones, known as sutures, separate the three major terranes: the Cameron fault zone in the Western Highlands and the Honey Hill fault zone in the Eastern Highlands (fig. 1–4, bottom). Because the latter turns rather abruptly northward and underlies a large lake in eastern Connecticut, geologists have given that segment a different, Indian name: the Lake Char, or Chargoggagoggmanchauggagogg-chaubunagungamaug, fault. Apparently, Indian anglers on opposite shores of this lake quarreled, which led to the following compromise: "You fish on your side, I fish on my side, and nobody fishes in the middle—no trouble."

2

Weather and Climate

Hurricanes and Ice Ages

I reverently believe that the Maker who made us all
Makes everything in New England but the Weather.
—Mark Twain

Connecticut is one of the smallest states in the Union. It measures about 125 miles from east to west and averages 65 miles from north to south. Twain, quoted above, also wrote, "As to the *size* of the weather in New England—lengthwise, I mean. It is utterly disproportioned to the size of that little county. . . . She can't hold a tenth part of her weather."

Connecticut weather is the result of many factors, all of which originate far beyond the state borders. Connecticut is located almost halfway between the equator and the North Pole and lies across the zone where cold, dry air from Canada's Arctic region alternates with, and frequently encounters, warm moist air moving north from the tropical Gulf of Mexico. These air masses jockey back and forth and at times cause great turbulence. The Atlantic Ocean has a modifying effect, but only for a relatively narrow belt of territory along the coast.

The state's weather changes are therefore rapid and often quite pronounced. Twain wrote: ". . . one of the brightest gems in the New England weather is the dazzling uncertainty of it." Along with the interplay of continental polar and maritime tropical air masses, the state's topography also plays its part. Significant snowfalls and plunging temperatures characterize the Northwestern Hills in winter, while the low-lying coastal plain boasts much milder conditions. The waters of Long Island Sound heat up slowly in the spring and stay warm longer in the fall. As a result, coastal summers are cooler and winters are warmer: when snow falls in Litchfield, residents of Fairfield carry umbrellas.

Ezra Stiles of Yale made the earliest scientifically recorded observations of Connecticut weather. He began his diary in June 1778 and continued uninterrupted until 1795.[1] After that year, others took over, which has resulted in a continuous record for more than two centuries. Mean annual summer and winter temperatures in New Haven between 1800 and 1900 are shown in figure 2–1.

In 1986, Michael Bell, author of *The Face of Connecticut,* wrote, "Connecticut is graced with a temperate, well-watered, forgiving climate." His statement contains much truth, but such weather extremes as blizzards, hurricanes, tornadoes, and floods are more common than most residents realize. On several occasions, nature has gone berserk, with calamitous consequences. Destructive hurricanes caused havoc in 1635, 1815, and 1938 (fig. 2–2). In 1985, Hurricane Gloria had the potential to join this triad but weakened just before hitting the New England coast.

The earliest hurricane recorded in New England was in fact the one that occurred on August 15, 1635, only fourteen years after the initial Thanksgiving celebration in Plymouth.[2] Judging by the damages described, it was probably a category three storm. In his history of the Plymouth settlement, William Bradford wrote: "It began in the morning, a little before day. . . . It blew down sundry houses, and uncovered divers others; divers vessels were lost at sea. . . . It caused the sea to swell in some places . . . so that it rose to twenty feet right up and down and made many of the Indians to climb into trees for safety. It threw down all corne . . . and blew down many hundred thousands of trees . . . the marks of it will remain for many years." The event could have been seen as a sign from God for these early settlers to return home across the sea, but that door was shut: many, if not most, of their ships had been badly damaged or destroyed as well.

1. The extremely cold winter of 1708–9 in Europe stimulated Gabriel Fahrenheit to design the first reliable thermometer, which Stiles used to his advantage.

2. An erosional level between two peat horizons in the coastal wetlands of Branford suggests heavy damage to the coastal zone caused by a hurricane sometime in the mid-fifteenth century. Further studies may reveal evidence of earlier events as well.

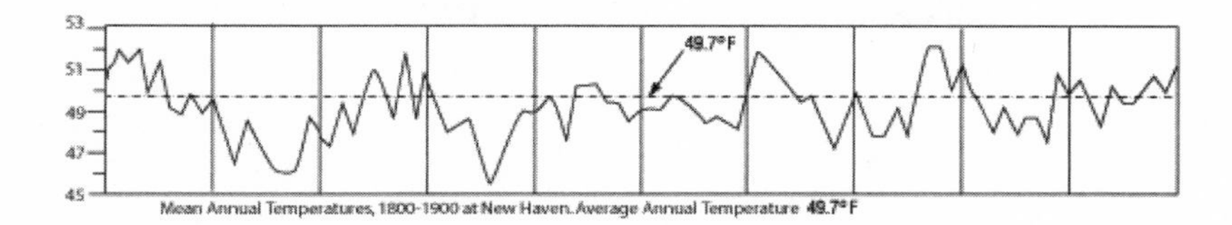

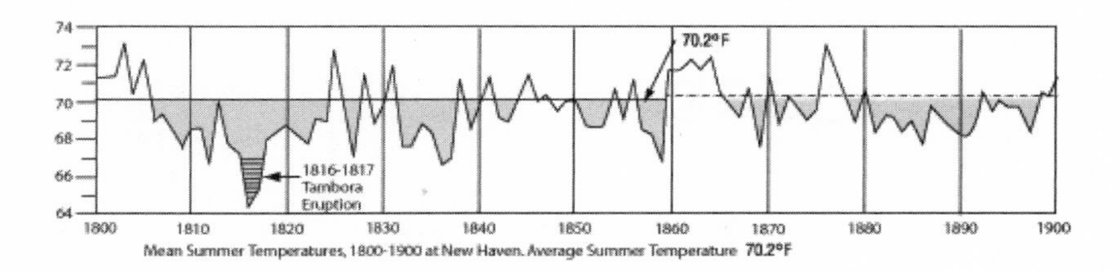

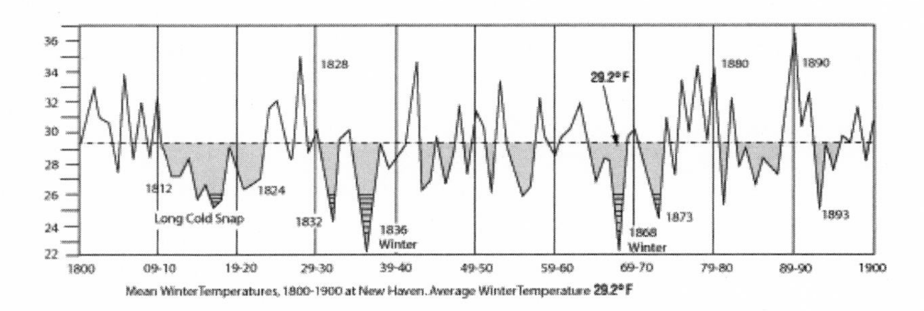

Fig. 2–1. Mean summer and winter temperatures in the nineteenth century in New Haven (Brumbach, 1965, figure 9). *Courtesy Connecticut Geological Survey.*

Another seventeenth-century hurricane brought heavy rains in July 1683 and caused floods in low-lying areas along the Connecticut River that proved detrimental to many in the colony. Increase Mather wrote: "Then on August 23 a second and more dreadful flood came: the waters were observed to rise 26 feet above their usual boundaries; the grass in the meadows, also the English grain, was carried away before it; the Indian corn, by the long continuance of the waters, is spoiled, so that the four river towns, viz. Windsor, Hartford, Weathersfield, Middle-Town, are extreme sufferers. They write from thence that some who had hundreds of bushels of corn in the morning, at night had not one peck left for their families to live on."

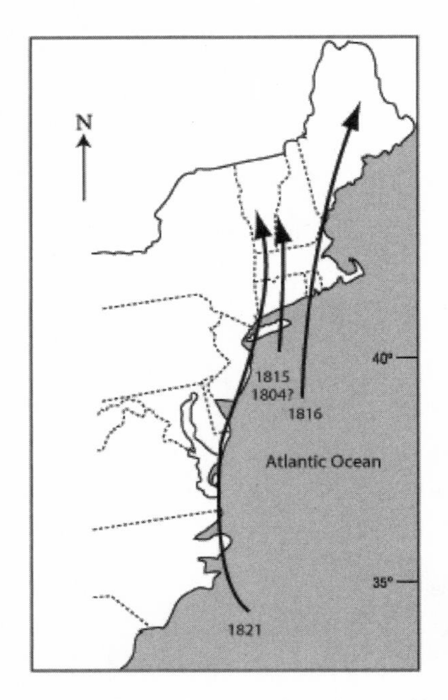

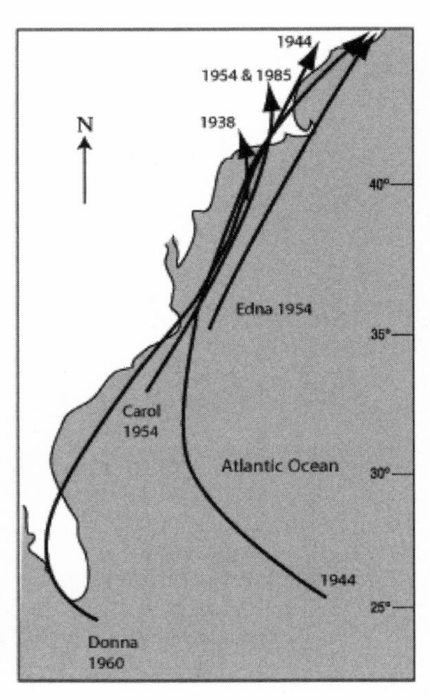

Fig. 2–2. Principal nineteenth- and twentieth-century hurricanes that affected New England *(adapted from Blair, 1992, figure 6).*

In the nineteenth century, more than a dozen hurricanes invaded the state or passed sufficiently close to it to cause damage. The strongest occurred in 1804, 1815, 1821, and 1869 (fig. 2–2). Timothy Dwight, president and professor at Yale, witnessed the great storm of October 1804 and reported that it broke the high pine trees and tall oaks in their "midst." The sight was "very strange and fearful to behold." Damage was most severe in the coastal zone between New London and New Bedford.

Victor Utley of Lyme wrote about the 1815 hurricane: "The morning was cloudy, with a cool wind and heavy rain from the northeast. . . . The violence of the wind increased gradually till 9 o'clock, at which time it blew a perfect hurricane and continued with the utmost fury until 11 a.m. . . . It tore down barns, unroofed dwelling-houses, upset cider mills, and carried away carriage houses. . . . It tore up the largest trees by the roots; some orchards are nearly destroyed. . . . The heart sickens at the sight. Whole forests of trees are either broken down or torn up by the roots and crossing each other."

Although Oliver Wendell Holmes was only seven in 1815, he remembered this traumatic event, especially because the wind "blew away his breeches," which were drying on a clothesline. Fifteen years later, he wrote "The September Gale."

> Lord! How the ponds and rivers boiled!
> They seemed like bursting craters!
> And oaks lay scattered on the ground
> As if they were p'taters;
> And all below was a howl,
> And all below a clatter,—
> The earth was like a frying pan,
> Or some such hissing matter.

William C. Redfield was the first to notice that hurricanes are cyclonic windstorms. From the directions of the trees blown down by the hurricane that entered Connecticut on September 3, 1821, he deduced that "this storm was exhibited in the form of a great whirlwind." The trees in central Connecticut had fallen to the northwest, while those in the northwestern part of the state and western Massachusetts had blown over to the southeast. A few years later, Redfield discussed his hypothesis with Professor Dennis Olmstead of Yale, who encouraged Redfield to record his observations. His essay, which appeared in the *American Journal of Science* in July 1831, was the first to discuss the dynamics of North American hurricanes and led to a better understanding of their destructive potential. Redfield, born in Middletown in 1789, began his career as apprentice to a harness maker and wound up a well-known meteorologist and first president of the Association of Naturalists and Geologists, which later evolved into the American Association for the Advancement of Science.

The worst twentieth-century hurricane, also known as the "New England Express," blew in on October 21, 1938 (see fig. 2–2). Three factors—gale, storm surge, and flooding—combined to bring about the greatest disaster in the history of southern New England. The hurricane originated as a tropical cyclone in the region of the Cape Verde Islands. Far out at sea, the approaching hurricane furrowed the ocean into giant windrows. Those swells preceded the storm's landfall by several days. They pounded the shores with such force that seismographs in California recorded the trembling. In twelve days, the storm crossed the Atlantic and curved north to hit southern New England head on. Although the hurricane had traveled some three thousand miles, its destructive effects were mainly confined to Long Island and the southern states of New England.

In *The Glory and the Dream,* William Manchester described the hurricane's landfall as a counterpart to the political storm blowing over Europe. He mentioned how Long Island served as a breakwater for Connecticut. Its manicured lawns a mile inland were under breakers two feet high. Long Island Sound, "beaten into one solid mass of foam," washed the dunes away, leveled beach houses, and stranded fishing craft. Without Long Island, Connecticut's coast would have experienced the same high waves that entered Narragansett Bay, washed across its islands, and left Providence under thirteen feet of water.

The eye of the storm entered between New Haven and Bridgeport and moved rapidly north. Meteorologists at the Blue Hill Observatory, near Boston, measured steady winds in excess of 120 miles per hour, and the gusts exceeded 186 miles per hour. Most ironic, given the rain, were the fires that consumed a quarter-square-mile area of the New London harbor district. The five-masted ship *Marsala,* which had broken free of its moorings, started the conflagration. Its bowsprit penetrated the brick wall of one of the harbor buildings when the ship was driven ashore, leading to an electric short circuit. Firefighters could do little as the winds that fed the fires gusted to ninety miles per hour.

Hundreds of stately elms and many white church spires, the pride of New England, crashed to the ground under the force of the storm. Wesleyan lost the steeple of its hundred-year-old chapel. Few town greens ever returned to their former splendor. Winds penetrated deep inland and carried sea salt that whitened windows even in Vermont. Factories in the mill towns were flooded, and the dams that provided their power were breached or severely damaged. Tides were high, causing storm surges ranging from nine to twelve feet above mean sea level. New beaches and flooded wetlands permanently altered Connecticut's shoreline. Hundreds of houses washed away. Heavy rains totaling seven to ten inches preceded the storm, and an additional three to seven inches fell as the storm crossed the state, resulting in extensive flooding. The Connecticut River at Hartford rose twenty feet above flood stage on October 2 and 3. More than seven hundred people died; more than sixty thousand lost their homes.

Six additional hurricanes blew in from the Atlantic during the twentieth century. An unnamed category three hurricane hit New England in 1944. It was followed by Carol in 1954, Diane in 1955, and Donna in 1960, then later by Agnes in 1972 and Gloria in 1985 (fig. 2–2).

Tornadoes have also occasionally visited the state. Since the first record of a tornado in 1682, at least forty major twisters have swept through. The widths of their destructive paths ranged from as little as ten yards to as much as several hundred yards; the lengths of the devastated zones varied from several hundred yards to more than ten miles. One notable tornado hit Connecticut on May 24, 1962, between 5:30 and 6:00 p.m. Its twelve-mile-long swath stretched from Middlebury through Waterbury and Wolcott to Southington. Because ample warnings from the U.S. Weather Bureau preceded the tornado, only one person died, but fifty were injured. In 1979, a tornado tore through Windsor, heavily damaging Bradley Airport and its museum. Three tornados touched down on July 10, 1989. The northernmost began its path of destruction in Cornwall, leveling one of the few virgin pine forests left in the state. It then continued southeast and damaged much of the village of Bantam before weakening. Soon thereafter, a second touched down in Watertown, passing through Oakville and the northern part of Waterbury and destroying more than one hundred homes. The third and most destructive touched down in Hamden. The damage was restricted to a path only five miles long, but in that stretch almost four hundred structures were damaged. Fortunately, the tornado stopped short of the much larger city of New Haven, where nevertheless wind gusts of eighty miles per hour were measured downtown.

Climate in the Eighteenth and Nineteenth Centuries

Early settlers found the New England climate harsher, and its storms more violent, than what they knew from the Old World. One frustrated colonist who spent several decades in Rhode Island described the climate as profoundly intemperate, with excessive heat and cold, sudden, violent weather changes, terrible and mischievous thunder and lightning, and unwholesome air. Legend has it that the "confounded weather" so disillusioned Connecticut colonists that they decided to change Weathersfield's original town name to Wethersfield!

Eighteenth-century colonist Cotton Mather believed, however, that the climate had actually improved since the earliest settlers' arrival in New England and offered an interesting hypothesis. "Our cold is much moderated since the opening and clearing of our woods." Benjamin Franklin himself agreed and provided an explanation for this "phenomenon" in a letter to Ezra Stiles dated May 29, 1763: "Snow lying on the Earth must contribute to cool and keep cold the Wind blowing over it. When a Country is clear'd of Woods, the Sun acts more strongly on the Face of the Earth. It warms the Earth more before Snows fall, and small Snows may often be soon melted by that Warmth. It melts great Snows sooner than they could be melted if they were shaded by the Trees."[3]

In 1809, Professor Samuel Williams demonstrated these effects scientifically by comparing soil temperatures in cleared pastures and their adjacent wooded lots, noting a difference of ten to eleven degrees Fahrenheit. Two years later, Hugh Williamson of Harvard College concluded that recent accumulations of snow in New England were less than half of what they had been fifty years earlier: "It is well known that in the Atlantic States, the cold of our winters is greatly moderated. As the surface of the country is cleared, a greater quantity of heat is reflected; the air becomes warmer, and the north-west winds are checked in their progress." He predicted that the changes would turn New England into "a proper nursery of genius, learning, industry and the liberal arts."

The Year without a Summer

The summer of 1816 was the coldest on record for almost two centuries (fig. 2–1, middle). In 1857, climatologist Lorin Blodget wrote, "In the northern states, snow and frosts occurred in every month of the summer.

3. Widespread deforestation of southern New England during the late seventeenth and early eighteenth centuries is clearly shown by the rapid increase of ragweed pollen in sediment cores.

Indian corn did not ripen; fruits and grains of every sort were greatly reduced in quantity or wholly out off." In Connecticut as much as three-quarters of the corn was unripe, moldy, and soft at harvest time. The loss of this common crop was a major setback, especially for isolated subsistence farmers in the Highlands whose survival depended entirely on nature's caprice.

On June 15, 1816, the *North Star*, a Vermont newspaper, published the following beneath the headline "Melancholy Weather": "From the 6th to the 10th [of June] it froze as hard five nights in succession as it usually does in December. On the night of the 6th water froze an inch thick and on the nights of the 7th and 8th it snowed, drifting locally to depths of 18 to 20 inches." A killing frost on June 10 defoliated large swaths of forest on the north-facing slopes of Connecticut's hills, creating a bleak midwinter scene in early summer. Calvin Mansfield of North Branford, Connecticut, wrote on June 11, "Great frost—we must learn to be humble."

Throughout Canada and New England, the abnormally cold weather and droughts devastated crops, including wheat and the all-important hay and corn that provided fodder for farm animals. In the Green Mountains, farmers collected the inner bark of white birches and milled it into flour. Those in the lowlands did not have to go to this extreme but still had to feed their pigs with fish caught in local streams for lack of fodder. Others had fish shipped in from nearby seaports, at high cost. Thus, 1816 is remembered not only as the "year without a summer" but also as the "mackerel year."

The farmers' most serious problems arose in the spring of 1817, of course, when there was little seed left for that year's crops. Because most farmers were self sufficient, little currency cycled through their communities so few people could manage to buy seeds. In addition, the price of wheat and other grains had risen from less than two dollars to more than three dollars a bushel.

From the very beginning of the nineteenth century onward, there were constant migrations west, because almost all of the fertile land in New England had been settled and families had in turn grown rapidly. The epitaph for Mary Buel, the wife of Deacon John Buel, who died on November 4, 1768, at the age of ninety, exemplifies the population explosion in the late eighteenth century: Mrs. Mary had "13 children,

101 Grand-children, 247 Grate-Grandchildren, and 49 Grate-Grate-Grandchildren"—410 descendants, of which 336 survived her. Because of food shortages and many hungry mouths, the trickle of New England farmers settling in Ohio—then known as western Connecticut—grew to a steady stream, especially in the years following 1817.[4]

That cold miserable summer and its crop failures and famine were not restricted to New England. In Europe, snow and frosts came early and stayed late, significantly shortening the growing season. A lack of seed and incessant rains frustrated attempts to replant summer wheat, and granaries throughout Europe emptied. The bad weather exacerbated the social and economic disruptions caused by Napoleon's wars. The scaling down of weapons factories and general demobilization had caused widespread unemployment, and most of those laid off lacked farming experience and could not replace the astounding number of young farmers who had perished in the fighting.

In France, rioting and pillaging broke out when a tax was levied on wheat, which had already doubled in price. The unrest spread across Europe and reached a climax in the spring and summer of 1817. These "grain riots" were marked by a level of violence exceeded only by the French Revolution. The "imminent end of the world" was prophesied amid outbursts such as Ireland's typhus epidemic, which persisted for several years. The number of deaths due to this outbreak alone has been estimated at more than 100,000. Many of its survivors in turn immigrated to North America.

The dreary, cold, and wet summer of 1816 even induced Lord Byron to write the poem "Darkness," a reflection of that melancholy time:

> I had a dream, which was not all a dream.
> The bright sun was extinguish'd, and the stars
> Did wander darkling in the eternal space,
> Rayless, and pathless, and the icy earth
> Swung blind and blackening in the moonless air;
> Morn came and went—and came, and
> Brought no day,

4. A report prepared at Ohio State University states that the first general wave of farmers to move from New England to the Midwest occurred after 1817.

> And men forget their passions in the dread
> Of this their desolation
> All earth was but one thought—and that was death,
> Immediate and inglorious, and the pang
> Of famine fed upon all entrails.

Byron wrote this poem at a villa he had rented on the shore of Lake Geneva.[5] His "famine" no doubt referred to the thousands of Swiss people who abandoned their highland farms and swarmed into the towns and cities, begging for food.

Although it is difficult to comprehend, the American and European cold snaps, famine, epidemics, and diasporas were all the result of a deterioration of global weather systems that had been caused by the eruption of a single volcano in Indonesia, many thousands of miles away. The Tambora volcano on the island of Sumbawa began its eruption cycle on April 5, 1815. Activity climaxed on April 10 with a series of prodigious blasts that shot gases and ashes into the stratosphere. The ground trembled and shock waves rattled houses more than five hundred miles away. Volcanic sulfur dioxide gases combined with water vapor to form veils of sulfuric aerosols that spread around the globe. Those veils, composed of tiny droplets of acid, reflected part of the sun's incoming rays back into space and thus reduced the amount of warmth reaching the surface, which caused major variations in global weather patterns and, in turn, the cold summers in 1816 and 1817. The debilitating effects of those years were most pronounced in the densely populated agrarian-based countries of Europe, where they brought social unrest and economic despair. Their impact on North America was somewhat less severe.

1826–27: The Year without a Winter

Extreme warm weather conditions also occurred in New England in the early nineteenth century. The spring of 1826 began with a severe

5. The British poet Percy Bysshe Shelley had rented a chalet nearby with his future wife, Mary. The cold and incessant rains limited them to sitting in front of their fireplace for days on end, during which time Mary wrote her immortal horror story, *Frankenstein.*

freshet that caused considerable flooding along the Connecticut River. It was followed by an intensely hot and dry summer that brought a plague of grasshoppers. Fall and early winter were unusually warm, so that flocks could remain out at pasture until January 8, 1827. Heavy snows then followed, but there was ultimately no frost in the ground. Farmers in New Hampshire were able to plow their fields by the end of February, even though snow was still covering the mountains. Unusually warm nineteenth-century summers returned from 1860 through 1865 and 1876 through 1878 as well.

Climate Cycles

The variations in weather patterns mentioned so far were all relatively short and do not in and of themselves constitute climate change. The Earth's climate in fact results from a rather complex interplay of external and internal factors over long periods.

In his famous 1920 publication on thermal phenomena, the Serbian mathematician and astronomer Milutin Milankovitch showed that the distribution and amount of solar energy reaching Earth varies over time. He concluded that Earth had experienced ice ages because its orbit around the sun is eccentric (elliptical), and the Earth during those periods was furthest removed from its source of heat. His hypothesis was initially received with much disbelief, but Milankovitch persevered in his research. Today orbital eccentricity has been generally accepted as one of the major parameters influencing climate.

The Earth's orbit varies from near circular to somewhat elliptical in cycles of 95,000, 123,000, and 413,000 years. The tilt of Earth's spin axis varies between 21.5 degrees and 24.5 degrees, with a periodicity of 41,000 years. In addition, the spin axis wobbles in a 19,000- to 23,000-year precession (fig. 2–3). All of these movements combined influence the amount of solar energy received at different times and on different parts of the planet's surface during its waltz through space. Climatologists refer to the 100,000-year eccentricity cycle as the "pacemaker" of the ice ages, because its effect on the Earth's climate appears to be the most pronounced and long lived. This is certainly true for the last 800,000 years, when ice sheets took about 90,000 years to grow and

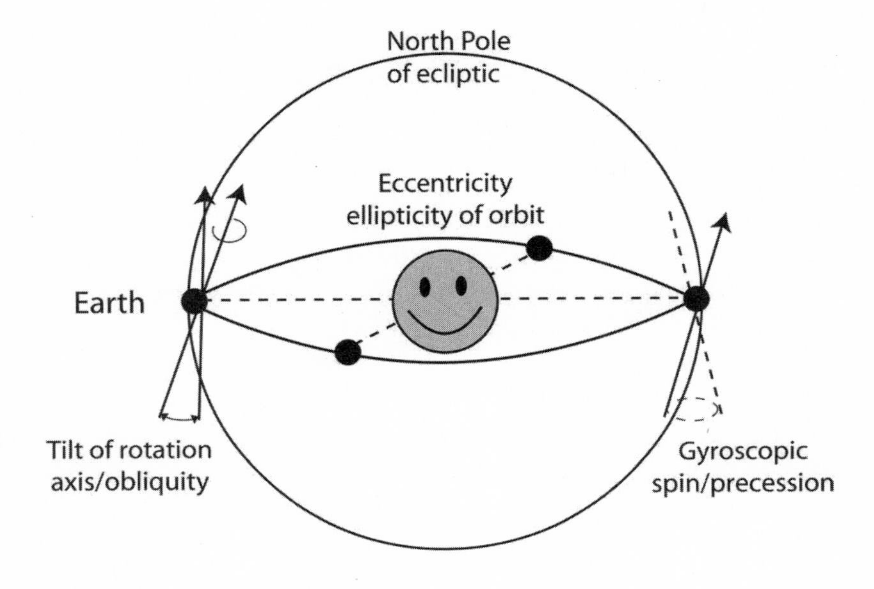

Periodicity; Eccentricity ----413.000 years
----100.000
Obliquity ------41.000
Precession -----23.000
-----18.000

Fig. 2–3. The movements of Earth in space that cause variations in the seasonal and geographic distribution of solar energy, thereby periodically changing the Earth's climate.

about 10,000 years to melt back, a periodicity that fits the cycles caused by changes in the Earth's orbit. Before that time, however, from about three million to about one million years ago, glacial cycles typically lasted about 41,000 years and matched instead changes in the Earth's tilt. The reason for the change in synchronization that occurred about one million years ago remains unclear.

To make things even more complex, non-cyclic cold periods occur as well. When large volumes of melt water that have accumulated in huge glacial lakes break through retaining dams and reach the oceans, they diminish thermohaline circulation—the currents controlled by differences in the temperature of seawater. A four-hundred-year cold spell, for instance, followed the emptying of glacial Lake Iroquois (the Lake Ontario basin) about 13,350 years ago.

Internal processes, both long- and short-term, also affect climate. The very slow drift of the continents, which are moved by convection currents inside the Earth's mantle, causes slow but significant climate change. In the last sixty million years, Greenland and Europe separated, forming the northern basin of the North Atlantic; India collided with Asia, causing the Himalayas to rise; and the Isthmus of Panama closed, separating the Atlantic and Pacific Oceans. All three movements had a global impact on oceanic and atmospheric current systems and so altered the climate.

Rapid climate changes of short duration lasting decades may also result indirectly from variations in the intensity of the magnetic field, which originates inside the outer core of the Earth; periods of major drought, for example, have been correlated with rapid increases in geomagnetic field intensity. The protective magnetic field that surrounds the Earth modulates the cosmic ray influx from the sun and influences the nucleation rate of clouds in the atmosphere. Cloud cover controls the Earth's *albedo* (the degree or extent to which its atmosphere reflects sunlight), surface temperatures, and rainfall, and thus its climate.

Three major periods of drought that could be related to magnetic events have occurred in North America in the last millennium. The earliest drought lasted from 1260 to about 1300 and caused the collapse of the Anasazi Indian culture at that time. The second and most severe drought spanned the two decades from 1572 to 1593; the dry conditions persisted into the early 1600s and extended eastward from the center of the continent as far as coastal Virginia, where they were most likely responsible for the disaster that awaited the earliest settlers at Roanoke and Jamestown. More than 80 percent of the colonists starved to death. The third drought stretched from 1910 to 1960 and included the infamous Dust Bowl, which lasted from 1933 to 1938. It contributed to one of the most devastating agricultural, economic, and social disasters in the history of the United States.

The Little Ice Age

The Little Ice Age was not a true ice age (fig. 2–4, top). Although mountain glaciers grew, covering farms and even destroying villages in mountainous areas, extensive ice sheets did not develop and changes in mean temperatures were relatively small at one to two degrees Fahrenheit. But

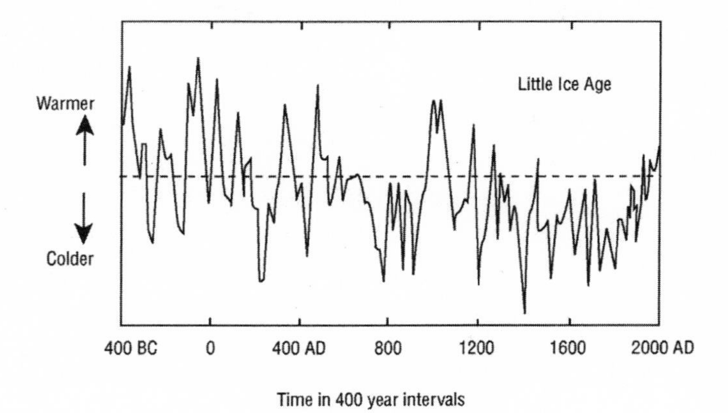

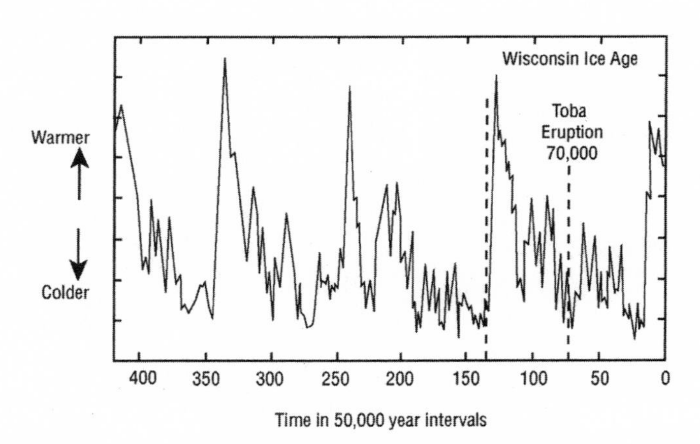

Fig. 2–4. Top: The Little Ice Age (adapted from McIntyre and McKitrick, 2003, figure 8). Bottom: The Wisconsinan and previous ice ages. The eruption of the Toba volcanic complex on Sumatra, which significantly worsened global climate, is believed to have occurred about 74,000 years ago. Antarctic ice core data *(adapted from Petit et al., 1999, figure 2).*

those minor fluctuations had major climatic consequences, changing the weather patterns and especially resulting in long, cold winters and significant variations in precipitation.

Its coldest periods shortened growing seasons in the northern hemisphere by three weeks and lowered the elevations at which crops could ripen by as much as five hundred feet, forcing numerous farms on mountain slopes to be abandoned.

During the Medieval Warm Period that preceded the Little Ice Age, western North America experienced widespread droughts; in Europe, winters were short and mild, summers long and dry. Vineyards in southern England produced almost as much wine as those in Bordeaux, and many British gentlemen became accustomed to a regular bottle of wine with their meals.[6] Then, around 1315, the European weather changed abruptly and a series of cold, wet summers ensued. Heavy rain fell relentlessly, turning farmland into wetlands. Whatever farmers planted soon rotted in the fields. Harvests in 1315 and the years immediately following were little more than half the annual yields of preceding years, which posed a major problem for a population that had grown rapidly during the Medieval Warm Period, and there was even less fodder for farm animals. The cold, torrential rains led to the deaths of hundreds of thousands of sheep throughout Europe, especially in England and France. Their absence in the fields resulted in decreasing amounts of manure, which in turn extended the years of meager crops that those fields produced. The first cold cycle of the Little Ice Age lasted only about seven years, but its social and environmental consequences continued for another decade, because the food shortages led to famine and then larger economic problems and the inevitable spread of disease.

In the following decades the weather patterns remained unpredictable. Violent storms flooded the lowlands of the northern Netherlands and Germany, bringing about the "Great Drowning" (1343–1362). During the same period, the Black Death devastated Europe's population; half of London's citizens died of bubonic plague in 1349 alone.

The weather improved somewhat in the next century, but by 1425 the cold had returned and became severe. Rivers and harbors that remained frozen for six months or longer characterized the winters in the second half of the fifteenth century. In the summers, fishing fleets were frequently caught in bad weather and could not deliver the protein (mainly herring) that many Europeans, who were otherwise living on bread and water, so badly needed.

6. A similar warm period is predicted for the decade from 2010 to 2020, and wine vineyards may be forced to migrate north or to higher elevations. In 1978, for example, the annual grape harvest in the French Alsace started on October 16; in 1998, it started on September 14; by 2007 the warm summers required harvesting to begin on August 24.

Because of the diversity of their resources and the low population densities, the native populations in New England most likely adapted reasonably well to the climate changes. There might have been local food shortages, however, especially among tribes that relied on corn staples. Political change sometimes ensued: ethnologists have suggested that famine led some Indian tribes to form leagues and temporarily set aside their differences at this time.

In the early part of the seventeenth century, the climate became somewhat better, and Dutch merchants decided to send explorers west to America to find "merchantable commodities." The coastal waters of New England teemed with fish; its streams abounded with beavers, whose furs provided warm coats for the wealthy; and its forests were filled with timber. Henry Hudson, in 1609, and Adrian Block, from 1610 to 1614, explored the coasts and principal rivers of future New York and Connecticut and established trading posts that became highly profitable.

Around 1670 the cold returned for a third time and was more severe than the cold snap at the end of the sixteenth century. This cycle lasted a little more than half a century (from 1670 to 1730 or so). New England winters were severe. The western segment of Long Island Sound frequently became a wide expanse of ice, allowing sleighs and sleds to cross from Fairfield to Long Island, a distance of about eighteen miles. The first snow of the season often arrived early in October and the last flakes fell well into May. Much has been written about the "Great Snow" of the winter of 1717; Joshua Hampsted of New London reported that "the snow is drove in some places 10 to 12 feet deep." In such winters, all life stood still while people huddled together in front of their fireplaces.

The careful recording of temperatures in New Haven from 1780 to 1968 permits a particularly deliberate plotting of winter and summer temperatures (fig. 2–1), and we see several periods of unusual cold. The longest cold snap of the nineteenth century in New England occurred between 1812 and 1824. In those years, snow covered the fields from mid-October to mid-May, and each winter the Connecticut River froze solidly. Farmers experienced significant hardship, but widespread famine does not appear to have occurred, as was the case in Europe. Connecticut farmers owned their land and were accustomed to setting aside ample reserves for following years, whereas in Europe tenants farmed the land. Their wages were at or below starvation level even when things went right, and they frequently could not store sufficient food for the winters. Many were forced to migrate to developing industrial complexes and cities, though conditions were not much better there either.

The cause of the Little Ice Age remains unclear, though a significant decline in sunspot activity appears to have occurred during those cold periods. The sun's output waxes and wanes in more or less regular cycles of about eleven years (the Schabe-cycle).[7] Since 1840, the maximum number of sunspots within each solar cycle has varied, from as few as sixty to as many as two hundred. The total number of sunspots between 1645 and 1715 was much smaller than the average number occurring in a single active year at present. This period of reduced sunspot activity became known as the Maunder Minimum, after the British astronomer who demonstrated a possible correlation between the variation in sunspot numbers and the Earth's climate.

Seventeenth-century scientists had noticed the lack of aurora borealis displays (the northern lights) in Europe's northern skies but had not linked them to sunspots. Those light shows are especially vivid when large numbers of subatomic particles, ejected into space by solar eruptions, enter the Earth's atmosphere near the poles and collide with particles in the ionosphere, which then "dance," full of excitement at receiving the sun's energy. When fewer dances were taking place, the Earth turned cold.

The Great Ice Age

Earth's climate began to turn colder some fifty million years ago. The change was more or less gradual, until a sharp drop occurred between thirty-two and thirty-four million years ago, right around the time when ice sheets began to form in Antarctica. After a relatively short rise in global temperatures between twenty-four and twenty million

7. There are also longer cycles of about twenty-two years (the Hale-cycle) and eighty to ninety years (the Gleisberg-cycle). Evidence for all three has been discovered in corals of the Galapagos Islands and sediments (varves) in glacial lakes. Unusually high sunspot activity has characterized the last seventy years and may have added to the global anthropogenic temperature increase.

years ago, the cooling trend continued, and the last five million years have been the coldest. Major ice sheets began to form around three million years ago in the northern hemisphere, on the continents surrounding the Arctic. Warmer and colder periods then ensued, each starting with a relatively slow drop in temperature and ending with rather abrupt warming again, which results in the characteristic "sawtooth" pattern of glacial and interglacial periods (fig. 2–4, bottom). This pattern can be best described as a climate with the hiccups. Global mean temperatures during glacial maxima are estimated to have on average been about nine degrees Fahrenheit lower than today's mean temperatures.

The most recent (Wisconsinan) ice age began about 120,000 years ago (fig. 2–4, bottom). Its ice sheets extended as far south as the fortieth parallel and covered two-thirds of the North American continent, including New England. Further south, savannas and deserts replaced forests, further reducing North America's biomass.

Major volcanic eruptions then temporarily aggravated the cold. About 74,000 years ago, a gigantic eruption of the Toba volcanic complex occurred on the Indonesian island of Sumatra. Widespread sheets of dust covered southern Asia and Africa, and dense clouds of sulfuric acid blanketed the entire Earth, causing a global "winter" that may have lasted from five to ten years (fig. 2–4, bottom). In the years following that eruption, temperatures are believed to have been near freezing in the subtropics, while hard freezes occurred in the middle latitudes, severely affecting the global ecology. This event may have in fact changed the course of human evolution. Based on studies of mitochondrial DNA, some scientists believe that only a few thousand early humans survived this global environmental disaster. The human race, then, barely escaped its extinction.

As ice masses thickened up north, their weight squeezed them downward and southward. Author Michael Bell compared it to "an enormous spill of molasses in a never ending January." Of the many ice sheets that covered northern North America, only two appear to have reached as far south as Connecticut, the first between 90,000 and 70,000 years ago, and the second between 35,000 and 23,000 years ago. A few dates suggest that the former may be much older and belongs to the Illinoian glacial event.

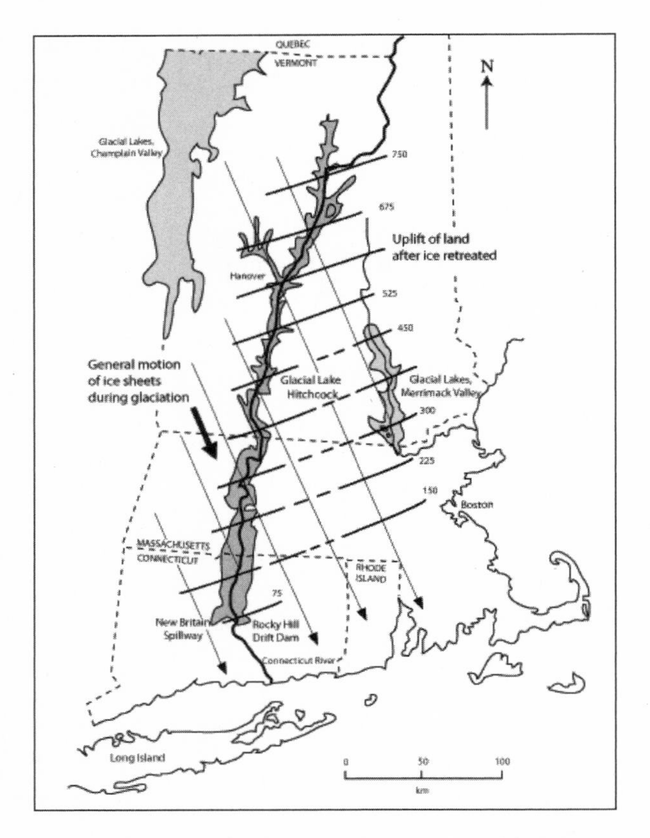

Fig. 2–5. Generalized south-southeast migration of the Wisconsinan ice sheet and extent of post-glacial Lake Hitchcock. Also shown: the differential uplift (in feet) of southwestern New England after the retreat of the ice (adapted from Koteff et al., 1988, figure 2).

While the earliest sheets of ice removed virtually all of the soil that had accumulated over many millions of years, the younger sheets ground the rocky ledges to bits, scratching and polishing the remaining surfaces in a slow but relentless process that created much of the present-day topography of Connecticut (fig. 2–6, bottom).[8] At the southernmost point where the melting of the sheets came to equal the supply of ice, the advance then stopped, and rocks and soils carried from the

8. Based on observations in New Hampshire, various geologists estimated that the last ice sheet in that state may have been as much as five thousand feet thick. By the time it reached Connecticut, however, it would have thinned considerably.

north were dumped into long, festoon-shaped terminal moraines. A well-known terminal moraine occurs along the northern shore of Long Island and stretches from Harbor Hill east to Orient Point, finally reappearing on Fishers Island.

About 20,000 years ago, the icy blanket slowly began to pull back. This glacial recession was not a smooth process, however; during several periods the ice margin stagnated and new moraines developed, forming narrow garlands farther and farther inland. The retreating ice front paused near Connecticut's shore about 20,000 years ago, leaving its remains in the piles of boulders that form a line stretching from Hammonasset (Meigs Point) to Ledyard. Farther north the ice, on at least one occasion, even readvanced southward for a short time.

On average, Connecticut's ice fronts retreated northward at a rate of about 150 to 250 feet per year. The melting ice blanketed the land with till—a chaotic layer of clays, sands, cobbles, and boulders—on which only thin soils have developed to date. Drumlins, streamlined whaleback-shaped hills of till with long axes that point in the direction from which the land ice came, emerged. Those elliptical hills, less than a mile long and rarely more than two hundred feet in height, became favorite sites for early settlements.

People walking the wooded areas and meadows of the Connecticut Highlands often encounter large, rounded boulders, called erratics, that had been plucked from resistant bedrock ledges, moved southward, and dumped on foreign ground when the ice melted (fig. 2–6, top). The largest known erratic in Connecticut is Cochegan Rock in Montville; its weight has been estimated at seven thousand tons.

As they went about the exhausting task of removing the erratics from their fields, early settlers had lots of time to consider their origins. Most attributed these out-of-place rocks to the biblical flood described in Genesis, but others at least attempted scientific explanations. During construction of his cotton mill in Vernon in 1825, Peter Dobson became curious about the markings on some of the erratics he had unearthed. In an 1826 letter to Yale professor Benjamin Silliman, Dobson wrote that he had dug up a great number of boulders composed of red sandstone with smooth undersides that showed scratches and furrows "as if done by their having been dragged over rocks and gravelly earth

South view of the Judges' Cave.

Fig. 2–6. Top: Judges Cave, large glacial erratic (~ 1000 tons), pushed onto West Rock during the Wisconsinan glacial period (*John Warner Barber, 1836, p. 152*). Bottom: Glacial striations on bedrock, Portland area.

in one steady position." He concluded that the boulders "have been worn by being suspended and carried in ice, over rocks and earth."

Dobson's hypothesis was the first that attributed these strange boulders strewn across New England to glacial activity. As often happens with truly original scientific ideas, his hypothesis received little attention at the time. Seventeen years later, however, Dobson's pioneering contribution to glacial geology was acknowledged in London by Sir Roderick Murchison, the president of the Royal Geological Society. In his 1843 address he congratulated the American sciences for hosting the original author of the best glacial theory, "though his name had escaped notice." Murchison concluded that an earlier acquaintance with Dobson's work might have saved numerous volumes of disputation on both sides of the Atlantic.

After news of the American "flood stones" reached Europe, several of its "wise men" felt compelled to weigh in. The Comte de Buffon, an eminent French geologist, attributed the responsible flood to "the malignancy of America's climate, the barrenness of its soils, and the imperfect nature of its animals and people." Timothy Dwight also heard from a "de Pauw" (Dutch for Peacock) who took it a step further, writing that the stones and marshes left by the flood were the cause of "America's insalubrities; the great number of its insects, the defectiveness of its quadrupeds, the barrenness of both the soil and its women and the stupidity of its men."

Glacial tills in the Highlands are especially rocky because the metamorphic rocks from which they derived were stronger and more resistant to grinding than the soft sandstone layers in the Central Valley. Odell Shepard wrote that "for more than 300 years, Connecticut's farmers dug rocks out of their fields, loaded them onto their oak sleighs and dragged them off toward their interminable stone walls." In *Stone by Stone,* Robert Thorson referred to one estimate of the total length of the stone walls in New England, an astounding 240,000 miles, further than the distance to the moon.

Plowing Highland fields was difficult, if not impossible, so most farmers cleared the land mainly for pasture. The deforestation caused sands and clays to wash downhill, however, continuously exposing more boulders. In addition, those just below the surface had the nasty habit of rising like mushrooms in the wintertime when the moisture below the stones froze and expanded, heaving up even large and very heavy boulders. Yale's Ezra Stiles frequently quoted a little rhyme when describing his family farm in the Highlands of Cornwall: "Nature out of her boundless store / Threw rocks together, and did no more."

During the last advance of the ice sheets, huge volumes of water accumulated as snow and ice on the northern continents and Antarctica, and sea levels dropped more than four hundred feet as a result, exposing large sections of the continental slope. Connecticut's territory doubled, and rivers had to travel much farther to reach the ocean, cutting steep canyons into the underlying sediment. When the ice melted, the sea level rose by as much as four hundred feet, drowning the land and its coastal river valleys. The buried valley of the Connecticut River near Saybrook, for example, is at least three hundred feet below present mean sea level and extends offshore below the continental shelf. In the previous three thousand years, sea level remained almost at the same level, but it has risen about eight inches since 1870.

Sea level continues to rise today. Satellite measurements have shown that it rose, on average, a little over a tenth of an inch per year in the last decade. Depending on how rapidly the Arctic and Antarctic ice masses are melting, this number could quickly increase to more than an inch per year, which would cause major problems.

During glacial maxima, the weight of the large ice caps caused the Earth's crust to bend down elastically. After the ice melted, the crust began to rebound but at a much slower rate than the melting itself. In northern Canada, where the ice sheet might have reached a thickness of nine thousand feet, the land continues to rise. This region has risen by only about half the amount needed to compensate for its former load of ice.

Released of its glacial burden rather earlier than these other regions, Connecticut has slowly returned to its pre-glacial elevation. The sea level began to rise about 19,000 years ago, and Connecticut's rebound began 3,000 years later. Although the rise of sea level slowed about 5,000 years ago, New England continued its slow rise. Inland regions, where the ice sheet was thickest and land depression greatest, have risen higher than coastal areas (fig. 2–5), tilting the land. The flat tops of the deltas that developed in glacial Lake Hitchcock, near the Massachusetts border, are about three hundred feet higher than those that were originally on the same lake level in the Rocky Hill area.

Evacuated by the ice, Connecticut remained a windblown wasteland for many thousands of years. The earliest vegetation appeared around fifteen to fourteen thousand years ago, consisting of dwarf birches, crowberry, sedges, and grasses. Over time, juniper, hazel, myrtle, and pines followed, and gradually Connecticut became habitable.

As the Earth continues its waltz around the Sun, the ice sheets will one day return to New England. In *The Skin of Our Teeth,* the playwright Thornton Wilder lays out our options in this regard:

Mrs. Antrobus [wondering about the unusually cold summer asks]: "What about the cold weather?"

Telegraph Boy [answers]: "Of course I don't know anything—but they say there's a wall of ice moving down from the North, that's what they say. We can't get Boston by telegraph, and they're burning pianos in Hartford. It moves everything in front of it, churches and post offices and city halls."

Mrs. Antrobus: "What are people doing about it?"

Telegraph Boy: "Well—uh—Talking, mostly." (Thornton Wilder, 1942)

Geologic Evidence of Ancient Climates

It is not possible to learn much about Connecticut's climate by studying its geology, because erosion has removed all of the sediments that might have been deposited in the 180 million years that preceded the glacial ages. However, some information about Connecticut's early Mesozoic climate (which lasted from about 230 to about 180 million years ago) can be deduced from the sediments that collected in the Central Valley. Remains of exposed basins similar to the Central Valley occur along the entire Appalachian chain from Nova Scotia to South Carolina. Farther south and east (offshore), rift basins are also present though hidden below blankets of younger sediments.

While some sandstone beds in Nova Scotia possess textures resembling desert dunes, indicating an arid climate, the coal seams in the rift zones of Virginia and North Carolina suggest a warm climate that was characterized by periods of relatively high humidity. Connecticut's valley, located between these extremes, contains evidence of alternating periods of both wet and dry conditions, represented respectively by black and russet-colored sediments (plate 4, top). The black layers were deposits on the bottom of lakes where iron was reduced because of the absence of oxygen. Under such anaerobic, rotting conditions, black iron sulfides formed. Those layers also contain biogenic material, part of which was transformed into oil and tar. During burial, the oil was squeezed out but the tar remained, intensifying the sediment's dark color.

The red beds were deposited on wide outwash plains (playas) that surrounded the lakes. Soon after the rains ceased, the sediments dried out and baked in the hot subtropical sun. Their iron was oxidized and turned into hematite (iron oxide), a blood red mineral.

Where a black unit overlays a sequence of red beds, the difference in color signals a climate change from dry to wet. The reverse applies to red beds overlaying a black sequence.

In the early 1980s, Paul Olsen, a professor at Columbia University, made a detailed study of these early Mesozoic sediments in the Newark rift valley of New Jersey. He measured the vertical distances between recurring units of black lake beds and completed a statistical study of the most common distances between those layers. The data provided five relevant groupings that were 19, 34, 83, 105, and 315 feet apart, respectively. Clearly, the lakes did not form randomly over time. Olsen believes that the intervals correspond to periods of 25,000, 40,000, 100,000, 133,000, and 400,000 years, respectively, and that the lakes developed in response to "orbital forcing." In other words, the waxing and waning of the lakes in the rift zones were due to climatic changes controlled by the movements of the Earth in space (fig. 2–3).

Red-black sequences with similar periodicities occur in the sediments of Connecticut's Central Valley. Drivers on the highways that cut east-west through the Metacomet ridge, where the Mesozoic sediments are exposed, can observe evidence of these climate variations, dating to many million years ago, by simply noting the color variations in the outcrops (plate 4, top).[9]

Climate changes related to the Earth's journey through space are intermittently overprinted by catastrophic events that cause relatively short but intense variations in climate patterns. The Paleozoic era ended and the Mesozoic era began about 250 million years ago, following one

9. One of the best exposures is along Route 9 in East Berlin, near the intersection with Route 15. There the red and black layers are visible in three dimensions, while their the fourth dimension, time, can be deduced from the layering.

of the most devastating ecological crises the Earth has experienced. At that time, emissions of huge volumes of basaltic magma in Siberia and their climatic aftereffects might have caused the extinction of numerous living organisms.

During the Triassic period, in the time of the Early Mesozoic, two significant but much smaller faunal mass extinctions occurred. In the Carnian period (about 216 million years ago), the losses appear to have been mostly on land; at the end of the Rhaetian period (about 200 million years ago), marine ecology received the greatest shock. The reason (or reasons) for the disappearance of much of the herbivorous archosaurus group of dinosaurs around 216 million years ago is unknown. It is possible that the extinction was related to the impact of large asteroids. The Manicouagan crater, in Quebec, has a diameter of seventy miles, and this impact would have caused days to turn into cold, dark nights for a period of many years, suppressing plant photosynthesis and causing mass starvation among animals. Unfortunately, the ages obtained for the Manicouagan impact vary, and this event could have occurred much later (about 214 million years ago) than the mid-Triassic extinction. However, it appears to be too much of a coincidence that a major impact and an extinction occurred so close together in geologic time,[10] especially when confidence limits in the dating procedures of old rocks vary by several million years.

The second major extinction occurred at the end of the Triassic period (about 200 million years ago) and has been used by paleontologists to separate the Triassic from the Jurassic period. This extinction coincided with enormous volcanic events along the seam between the North American and African continents during the initial stages of their separation, which led to the opening of the Atlantic basin. The region involved with this early Jurassic volcanism spread out over an astounding four million square miles and involved the margins of two entire continents.

Using the evidence for recurring dry and wet climates in the sediments between lava flows, Olsen estimated the time interval between the oldest and youngest of the three volcanic events in Connecticut to have been about 600,000 years. The periods of weather deterioration

10. Two additional asteroid impacts occurred in the same general period: one in western Canada (Saint Martin) and the other in France (Rochechouart).

associated with the cooling and degassing of the three volcanic events were therefore separated by several hundred thousand years, which left sufficient time for nature to regenerate and reestablish itself. However, many of the volcanic rocks were emitted along the outermost margins of the North American and African plates, and these events were much more voluminous. Little doubt exists that these gigantic emissions of magma caused periods of severe climate deterioration that may have lasted for decades, possibly centuries, to disastrous effect.[11]

Connecticut's Mesozoic sediments and volcanic rocks in its Central Valley thus provide clear evidence for the complex interplay of external (Milankovitch cycles) and internal (voluminous magmatism) phenomena that has resulted in significant climate changes and their impact on life forms.

Sedimentary formations in Connecticut that are older than 230 million years probably carried similar records of past climates, but their subsequent burial and deformation erased all of the evidence. Their fossils were destroyed, their colors changed, and their textures transformed by recrystallization. It is necessary to cross state borders in order to find out what Connecticut's climate may have been like in the Paleozoic. The Narragansett basin of Rhode Island contains late Carboniferous deposits that were little changed in the time since, and in the Catskills of New York, thick piles of Devonian and Silurian sediments are exposed and brimming with marine fossils.

Information on Earth's climate in the last 500 million years has been gathered by studying changes in the oxygen and carbon isotope ratios in the fossil remains of organisms that lived in oceans (plate 4, bottom). The most reliable part of the curve applies to the most recent 100 million years, because fossils of that age span can be collected from well-dated deep-sea sediments. Older ocean deposits exist mainly above sea level, pushed up onto continents during periods of mountain building. Those older fossils may have been contaminated by reactions with ground water, rendering the data less reliable though no

11. A similar sequence of voluminous volcanic eruptions occurred in the northernmost Atlantic basin during the separation of North America (Greenland) and Eurasia (Western Europe) about 56 million years ago. More than two hundred ash layers have been found in the Paleocene sediments of Denmark and in the North Sea that were probably derived from large volcanic centers in the region of present-day Iceland. Faunal and floral extinctions occurred in that period as well.

less remarkable. Of great interest is the apparent periodicity of about 140 million years between successive periods with Ice Ages (plate 4, bottom). Some scientists believe that this apparent cycle could be related to the movement of our solar system in its galaxy. Earth's encounter with certain galactic spiral arms could lead to an increase in cosmic ray influx, which would affect the atmosphere globally and change the climate. Others believe that the internal forces that shift the positions of the continental masses drove the changes instead.

The youngest (Pleistocene) ice ages occurred when Antarctica covered the South Pole, and the Eurasian and North American plates crowded around the North Pole. Although evidence exists for an earlier cold period, the Earth does not appear to have experienced an obvious ice age around 150 million years ago, when the continents were spread along the equator. In this case, internal processes may have ameliorated the effects of external ones. Maybe ice ages occur only when continental masses cover or surround the polar regions.

The most amazing aspect of our climate is certainly its complexity. It appears to be controlled by both internal and external processes with widely varying periodicities. Somehow humans are adding to this complexity by rapidly burning the Earth's energy resources and producing changes within a much shorter period than those of most naturally occurring climatic events. How the Earth will react remains the biggest unknown.

3

Connecticut's Geologic Treasures

Gems and Ores

Why seek foreign shores
for precious ores?
To me the case is clear
we need not roam
at all from home!
We've lots of "owers" here.
—Anonymous, 1764

New Hampshire comes to mind when we think about granite, California for gold, and Colombia for emeralds. However, Connecticut's Stony Creek was once a major producer of granite, the area around the town of Cobalt yielded gold, and Middlesex quarries provided gemstones. An amazing variety of rocks and minerals occurs in Connecticut, and the remains of quarries can be found almost everywhere. Most of the enterprises that were developed to extract these "riches," however, were not economically viable. In many cases, bankruptcies brought economic gains for their fraudulent owners only because they kept a large part of others' investments.

The historian Charles Harte wrote that if you had a hole and claimed a valuable mineral deposit within it, all you had to do was organize a company in order "to have the otherwise staid and sedate citizens fall over each other in their rush to subscribe to the venture." Some Nutmeggers appear to have been quite adroit in attracting such funds.

One farmer even used a shotgun loaded with native copper from the Bristol mine to "seed" the walls of his hole, dug near Cheshire.

Despite a long list of failed ventures in this regard, however, Connecticut did provide many valuable rocks and minerals that significantly aided its economic development (fig. I-2). The Salisbury iron deposits brought great wealth to the northwestern part of the state, transforming a wilderness into a major industrial complex that played a crucial role in both the Revolutionary and Civil Wars. Copper ores discovered near Bristol propelled a flourishing brass industry, and barite mined near Cheshire benefited New Haven's economy. Portland's brownstone quarries gained a national reputation, and builders used this stone to construct the famous mansions and row houses of New York City. It was eventually exported to almost every city along the eastern seaboard, and several loads were even shipped to San Francisco, by way of Cape Horn, to construct the mansion of a railroad baron. Granite quarries, worked primarily for gravestones and monuments, developed along the coast from Westerly to Greenwich and provided an extra source of income for towns with declining fishing industries.

The history of these enterprises, large and small, is a long record of efforts both richly rewarded and disastrously misdirected. Hope ebbed and flowed as fortunes were made and lost, and lives were sacrificed in the deep quarries and dark mines, and in the offices of desolate investors.

Governor Winthrop's Gold Ring and the Connecticut Charter

While hunting or exploring the woods surrounding their lands, colonists occasionally discovered useful rocks and minerals. Local street names in villages all over Connecticut attest to many sites where people found (or thought they had found) deposits of lead, silver, tin, and other ores. The most interesting among the latter discoveries is the gold that John Winthrop the Younger supposedly found near present-day Cobalt, and its apparent role in helping him procure a charter from King Charles II of England.

Historian Robert Black, who wrote *The Younger John Winthrop*, reconstructed this eventful day in May 1662: "One can imagine the scene: the sumptuous council chamber, the glittering council, perhaps the king in person, easy and affable for all his over-adornment, and before

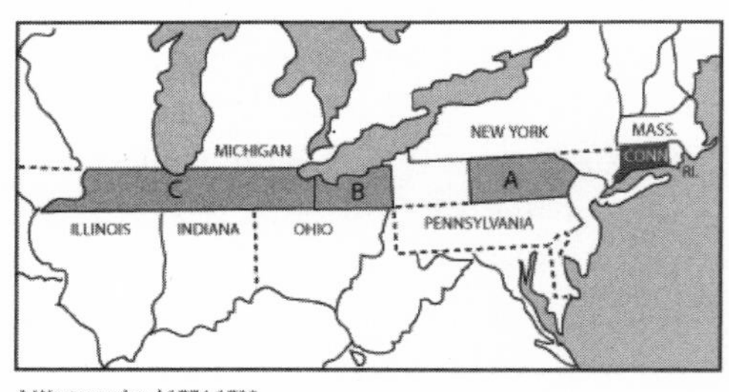

A Westmoreland 1774-1782
B Western Reserve 1786-1800
C Connecticut lands ceded to Congress in 1786

Fig. 3–1. Top: John Winthrop Jr., Connecticut's first governor. Bottom: part of the territory that, according to the 1662 state charter, belonged to Connecticut. Westmoreland was lost to Pennsylvania and the remainder, with exception of Ohio, ceded to the U.S. Congress.

them the governor of Connecticut quietly but suitably garbed, respectful but never for an instant a sycophant, his homely face aglow, his presentation couched in precisely the right phrases. It was the supreme test of his extraordinary personality, and it did not fail him: petition and charter draft were graciously received." It passed the seals, or was officially signed, on April 20.

The charter enormously magnified the territory of Connecticut's tiny original confederation of three towns, stating that its boundaries were to run "on the East by Norrogansett River commonly called Norrogansett Bay . . . and on the North by the Lyne of the Massachusetts Plantations and on the South by the Sea, and . . . from the said Norrogansett Bay on the East to the South Sea on the West Parte with all the Islands therevnto adjoyneinge." The "South Sea" in this case referred to the Pacific Ocean. Connecticut thus could lay claim to a huge slice of territory between the forty-first and forty-second parallels, extending across the continent from ocean to ocean. Figure 3–1 (bottom) shows the eastern segment of this belt.

Why had Charles II been so generous? According to a popular legend, Winthrop had won the king's favor by presenting him with a ring that he had supposedly forged from gold found near Great Hill, in an area near present-day Portland then known as Chatham. As the saying goes, "Gold then is gold, if stampt it is coyns. If fram'd in its due form, it becomes a Ring." Perhaps Winthrop gave the council the impression that more gold could be found in that area, which would have helped the British treasury—the king was to receive one-fifth of all "gold and silver" mined in the colony.

In 1754, almost a century later, commissioners representing several colonies met at Albany and confirmed that Connecticut, by right of its charter, extended west to the Pacific. Soon thereafter, Pennsylvania expropriated Westmoreland, a segment of the strip, despite the fact that the Connecticut charter antedated William Penn's by nineteen years. Connecticut ceded much of the remaining land to Congress in 1786 but retained its Western Reserve (present-day Ohio). Odell Shepard concluded that the Connecticut citizenry had not been too concerned about the loss of its westernmost territories. "Mount Shasta is indeed very mountainous, and the Great Salt Lake is extremely salty . . . neither one of them would harmonize with Connecticut's landscape."

John Winthrop Jr. was born in England in 1606 and immigrated to Boston in 1631. With the support of his father, he founded Ipswich (now Agawam, Massachusetts) and later moved south to found the town that would become New London. Connecticut suited him. He became magistrate in 1651, served as lieutenant governor from 1657 to 1658, and was then elected governor in each of the following seventeen years until his death in 1676. Well educated, Winthrop developed a profound understanding of the sciences, specializing in alchemy and medicine, and was elected to membership in the Royal Society. From early in his career, Winthrop recognized that the exploitation of New England's natural resources was essential to its economic development, and he was instrumental in establishing the ironworks at Braintree and Lynn in Massachusetts. In 1641, Winthrop journeyed to England and spent much of his time persuading merchants to invest in those iron-works. Thus Winthrop became the first metallurgist, prospector, and mining stock promoter in America.

In 1651, only nine days after Winthrop received his commission as magistrate from the Connecticut Court of Elections, the Connecticut General Assembly approved a generous contract that gave him unlimited mineral rights for developing mines of "Lead, Copper or Tin, or any minerals: such as Antimony, Vitriol, Black Lead (graphite), Alum, Salt, Salt springs or any other the like." Upon development, he would "enjoy forever said mines with the lands, woods, timber, and water within two or three miles of such mines." The order in which the metals and minerals are mentioned suggests their importance to Connecticut colonists: lead for bullets, copper and tin for utensils, and finally food additives such as salt.

Ten years later, on May 25, 1661, "the inhabitants of Midletowne, for ye encouragement of ye designes of our much honoured Governor, Mr. Jon Winthrop, for ye discovery of mines and mineralls, and for ye setting up of such works as shall be needfull for ye improvement of them, doe hereby grant unto our said much honoured Governor, any profitable mines, or mineralls, yt he shall finde or discover." Winthrop also obtained water rights and the use of up to one thousand acres of woodland to procure timber for refining eventual ores. In form and function, the Middletown grant resembled the 1651 grant. It did not state, however, that Winthrop was "to enjoy forever said mines"; the citizens of Middletown were stingier and set a development deadline of five years.

The only historic evidence that exists concerning Winthrop's search for minerals in the Middletown region comes from the diary of Yale president Ezra Stiles. On June 1, 1787, Stiles wrote, "Governor Trumbull has often told me that this [Portland, formerly Chatham] was the Place to which Governor Winthrop of New London used to resort with his Servant; and after spending three weeks in the Woods of this Mountain [Great Hill] in roasting Ores and assaying Metals by casting gold Rings, he used to return home to New London with plenty of Gold. Hence this [Great Hill] is called Governor Winthrop's Ring to this day."

When the Appalachian Mountains took shape, a massive rock formation was squeezed up toward the surface, producing Great Hill. This northeast-trending ridge rises to 740 feet and constitutes the border between East Hampton and Portland. It is composed mainly of erosion-resistant quartzite, which arose from metamorphosed beach sands. The more than 400-million-year-old sands probably contained gold that was remobilized during burial, when the sand turned first into sandstone and then into quartzite. In *Sand and Clouds* Annie Dillard explains that particular geologic process quite eloquently: "[Continental] Plates subduct. They tilt as if stricken and dive under the crust. At abyssal depths Earth's weight presses out their water; heat and pressure burst their molecules and sandstone changes into quartzite." High temperatures and pressures mobilized the gold and concentrated it, together with quartz, into thin veins.

In the three-week period Stiles refers to, Winthrop could have located a mineral deposit but obviously not developed it. Therefore, the gold must have been relatively easy to obtain, which suggests a "placer deposit," a concentration of grains of gold in stream sediments. A small, rapidly flowing stream, aptly known as Mine Brook, runs over bedrock parallel to Great Hill along its eastern flank. The stream bottom consists of a stepwise series of small falls and pools. Over time, water eroded the bedrock and its enclosed mineralized veins, liberating the gold from its imprisonment by the quartz. Because gold is many times heavier than quartz, its grains concentrate on the bottoms of pools. Winthrop needed only to order his servant to "pan" the sediment in the stream, and in a short period he would have collected sufficient gold to fashion one or more rings.

Great Hill became the focus of intensive mineral prospecting and mining in the eighteenth and nineteenth centuries (fig. 3–2). In the same paragraph in which he writes about Winthrop's gold, Stiles mentions a visit by Gominus Erkelens, who provided some details about his cobalt mine in the Great Hill area. Cobalt is a silvery ore that obtained its name from the German word *kobolds*, which refers to the mischievous ancestors of gnomes who live underground and take special delight in deceiving those mortals who invade their territory in search of Earth's riches. They play a role in Goethe's *Faust* and Mendelssohn's *Midsummer Night's Dream*.

Great Hill illustrates the suitability of this elusive metal's name. Dr. John Sebastian Stephauney, who had a horizontal opening carved into

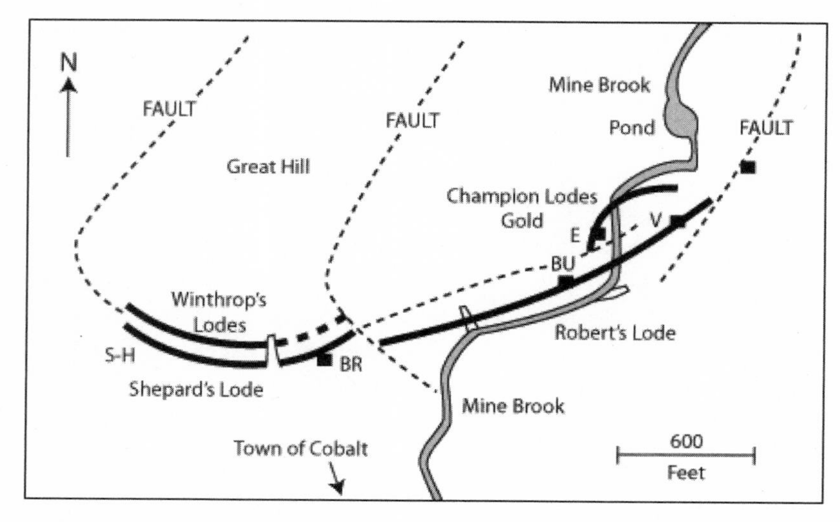

Fig. 3–2. Location of principal mineral veins (lodes) at the southern end of Great Hill, near Cobalt. SH: Stephauney-Hunt workings; BR: Browns shaft; BU: Bucks shaft; E: Engine shaft (~120 feet); V: Ventilation shaft (*adapted from Chomiak, 1989, figure 6*).

Great Hill to expose mineralized veins in 1762, woke the gnomes by starting actual mining in the Chatham area. Stephauney most likely had heard rumors about Winthrop's gold and decided to try his luck. After losing much of his money, he gave up, but he renewed his efforts a decade later and hired John Knool and Gominus Erkelens to help things along. Those fellows, presumably of Dutch or German descent, and trained in mineral prospecting, eventually discovered the presence of cobalt in the veins.

In 1775, Erkelens bought the mining rights from Stephauney, along the way constructing a distillery in Middle Haddam "to supply his workmen more conveniently with a beverage which they loved." Erkelens told Stiles in 1787 that he had collected as much as twenty tons of cobalt ore that he hoped to sell in China. Combining cobalt with flint and potash produces smalt, the compound that provides the deep blue and purple color in chinaware (plate 5, top). Imported porcelain with oriental designs was in great demand among New England's well-to-do citizenry in the eighteenth century, and exporting cobalt could have led to more balanced commerce with China. At their meeting, Stiles probably inquired after Winthrop's gold, but Erkelens, who had worked the mines for more than a decade, apparently remained silent on that subject. Erkelens and his load of cobalt ore disappear from history after 1787.

The saga of mining at the Chatham site continued in the nineteenth century. In 1818, Seth Hunt, a wealthy New Hampshire citizen, obtained the mineral rights and reopened the mine. Anticipating great results, he spent liberally, soon exhausting his own resources. He then formed a corporation with $20,000 capital. One of his more skeptical investors remarked, "Although Mr. Hunt read books on mining, he did not well understand his work." The investor also doubted that the men on whom the director relied for chemical and scientific information knew much more than Hunt did. Hunt's workers did collect as much as 1,000 pounds of cobalt ore. When samples were analyzed, however, the ore turned out to contain more nickel than cobalt, which made its refinement difficult and expensive. The mine closed three years later, in 1821.

The lure of riches continued nonetheless. In 1827, a group of investors from Philadelphia hired Edmund Brown and paid to sink a vertical shaft to a depth of thirty-eight feet with a drift extending some sixty feet eastward (fig. 3–2). Meanwhile, Brown erected a crushing plant, smelter, and laboratory for chemical analyses. Alarmed by such heavy expenditures, the investors sent Eugene Francfort to investigate. Francfort, a Middletown doctor, chemist, and self-proclaimed mining engineer, believed that richer veins existed that Brown's workers had somehow missed. He suggested further investments, but in 1829, the enterprise collapsed after little more than two years of mining activities.

Francfort attributed the failure to bad management and the introduction of newly invented, unproven machines. About the managers, he wrote, "Although they may [have been] excellent financiers and lawyers . . . the men were utterly unacquainted with the nature of minerals." He referred to the new crushing machines as "Iron Nutmegs" because they were made out of iron, but they could not even break Connecticut's famous nutmegs and were no better than those manufactured from wood. He continued to believe in the potential of the Great Hill site, and some of the minerals he had collected during his mining were so rare and beautifully crystallized that they were displayed at the International New York Crystal Palace Exhibition of 1853–54. They included cobalt bloom, nickeliferous mispickel, and finely disseminated smaltine (mineralogists can be rather fanciful when naming their minerals).

Those minerals drew the attention of a second group of investors from Philadelphia, who organized the Chatham Cobalt Mining Company in 1853 with an authorized capital of $500,000. They were primarily interested in nickel, a major constituent in coins that was in much demand. Francfort was hired as a consultant, and he joined Professor James Curtis Booth, an employee of the United States Mint, in providing several highly optimistic reports to the investors. Booth believed that the mines, when developed, would exercise a "powerful influence on the market for Cobalt and Nickel." Both men were wrong. The separation and refinement of these metals continued to be very costly, and the mining sputtered at best until 1859, when the site was abandoned for the last time.

Between 1762 and 1859, then, there were six intensive but brief attempts at mining the cobalt ores. In each case, the *kobolds* won and financial backers lost—while the miners apparently cut right through veins containing gold.

It is surprising that no one thought of using the cobalt locally. Feldspar, or "China stone," was quarried quite successfully throughout the nineteenth century in nearby Collins Hill, about two miles west of the cobalt mine, and in the hills on either side of the Connecticut River farther south. This mineral is used to produce porcelain glazes (plate 5, top). In 1837 alone, Middletown exported six hundred tons of feldspar to Liverpool, England, and one hundred tons to a porcelain factory in Jersey City. Had the cobalt been kept at home, Middletown might have become the Delft of the New World; chinaware with New Haven painter George Henry Durrie's (chapter 5) famous designs of New England winter scenes in blue, such as those shown on Currier and Ives prints, could have swept the nation and brought even more renown to Connecticut than its clock and gun enterprises did.

One hundred twenty-seven years after miners had ceased torturing the Great Hill area with picks, shovels, and explosives, their mines all long abandoned, faculty members and students from the University of Connecticut decided to study the tailings left behind. They collected samples of the rock and used a diamond saw to slice them up. While at work, they noticed curious yellowish streaks on the steel saw blade. Analyses confirmed it as gold and showed that one of the rock samples contained more than one hundred times the amount of gold commonly encountered in North American ores (plate 5, bottom).

Newspaper headlines read "Hills of Cobalt Hide a Real Gold Mine" and "Geologist Strikes Gold." The gold rush was on, as Stiles's story about Winthrop's gold suddenly changed from legend to probable truth. Many individuals and several companies inquired about the potential for reopening the mine, but because of its location in a state forest, it quickly became apparent that there would be too many legal, environmental, and zoning roadblocks. Interest faded almost as fast as it had arisen.

A question remains, however: How could all of the prospectors and miners who worked the site have missed the presence of gold? The veins also contain pyrite, known as fool's gold. Could the workers have mistaken one for the other? This could have been the case in the eighteenth century, but those who sampled and analyzed the ores in the nineteenth century were well-trained chemists and mineralogists who would have known the difference. It is instead possible, even somewhat likely, in view of the gold in the tailings, that the miners did recover it but kept it hidden from their supervisors and investors.

This mystery adds further interest to the legend of Winthrop's gold ring and the story of how Connecticut obtained its favorable charter from Charles II, winding up with a territory that greatly exceeded the size of the original colony.

Captain Jeremiah Wadsworth's Lead Bullets

Gold brought Connecticut its charter, but lead helped win its freedom. When the Revolution started, American soldiers desperately needed lead, and a vein near Middletown provided part of the ammunition used during the battles around Boston.

While exploring the Middletown area, in fact, Governor Winthrop might have located a lead vein exposed in Butler's Creek, a small stream joining the Connecticut River in the Maromas section of Middletown. He may have done so by using a method described in 1639 in *Subterranean Treasures* by Gabriel Plattes, who wrote that he had discovered a lead vein by cutting up a rod of hazel, about a yard long, and had wandered about until the wand bent downward. The only evidence that suggests that Winthrop found lead near Middletown consists of a few samples of galena, a lead sulfide, that were included in Winthrop's collection of Connecticut minerals sent to the Royal Society of Great Britain by his grandson. The evidence is strengthened, however, by the presence in the same collection of several specimens of feldspar and columbite[1] from the White Rock pegmatite outcrops a few hundred yards southeast of the future lead mine, buttressing the notion that Winthrop had collected in the area. Middletown lead occurs as galena in veins that consist primarily of quartz (plate 6, bottom). Because silver and lead are similar in atomic size and charge, the former often takes the place of the latter in the cubic galena crystals. Native silver was also present as separate veinlets inside the Middletown ore body; the silver content reputedly varied from ten to twenty ounces per ton of ore.

When war broke out between Great Britain and its American colonies, the rebels found themselves with sufficient light weapons but few cannons and a shortage of ammunition. They needed lead for their bullets. In April 1775, only days after the outbreak of hostilities in Lexington and Concord, a group of Connecticut conspirators, headed by rebels Silas Deane and Samuel Holden Parsons, took money from the colonial treasury—they did not have the consent of the Assembly, but they left promissory notes—to finance a secret raid on the British fort at Ticonderoga. The principal reason for the raid was to obtain cannons and munitions. The attack was successful, and the Americans "captured" 120 cannons, ten tons of musket balls, three cartloads of flint, ten casks of powder, and various weapons.

On May 10, 1775, Henry Knox took from that catch fifty-five cannons, 2,500 pounds of lead, and a barrel of flints and transported this bounty to Cambridge on ox sleds. The cannons proved to be a major incentive for the British evacuation of Boston in March 1776; much of the 2,500 pounds of lead was probably used to cast cannon balls. As the war progressed, the need for lead and powder continued to be urgent. Lead deposits, however, are scarce in New England; they occur mainly in relatively narrow quartz veins emplaced within early Mesozoic fault zones, such as those exposed near Middletown.

1. These columbite ($[Fe, Mn] Nb_2O_6$) samples became famous when it was discovered in 1802 that this mineral contained a then unknown element that was initially named columbium and later renamed niobium.

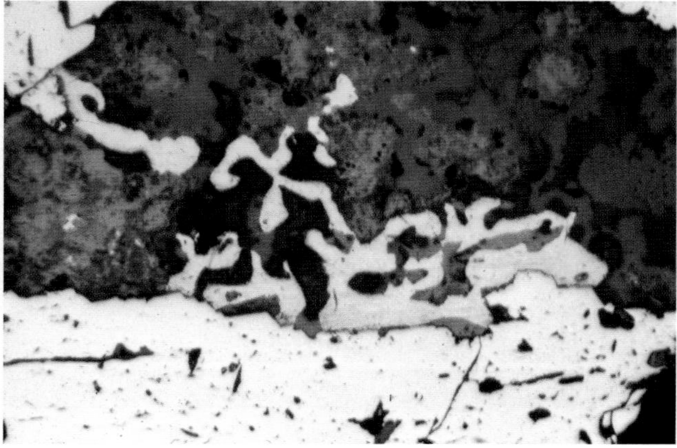

Plate 5. Top: Nineteenth-century China plate with cobalt blue design excavated in Middletown, and a sample of cobalt/nickel ore. *J. W. Peoples Museum, Wesleyan University.* Bottom: Polished thin section with gold in quartz matrix from Cobalt area. Quartz (grey-brown), arsenopyrite (white patches). *Courtesy Tony Philpotts, Yale University.*

Plate 6. Top: Southward view of the Middletown Silver-Lead mine (Bylaws of the Silver-Lead Mining Corporation, 1853). Bottom: Cubic crystal of galena (lead sulfide) in quartz matrix (Middletown Lead-Silver mine). *J. W. Peoples Museum, Wesleyan University.*

Plate 7. Top: Gem-quality beryl (golden heliodor) from East Haddam. *J. W. Peoples Museum, Wesleyan University.* Bottom: Six-sided beryl crystal in quartz matrix from the Gillette Quarry, East Haddam. *J. W. Peoples Museum, Wesleyan University.*

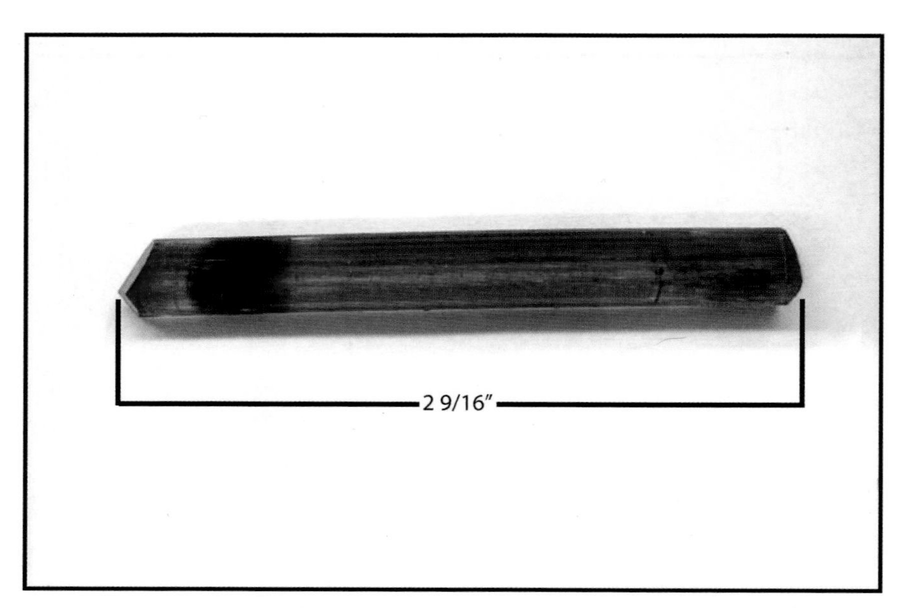

Plate 8. Top: Green tourmaline crystal with pink nose, Haddam area. *Courtesy American Museum of Natural History, New York, New York.* Bottom: Black tourmaline crystal in pegmatite matrix, Haddam area. *J. W. Peoples Museum, Wesleyan University.*

By the middle of the eighteenth century, "foreigners," notably British investors, had expended large sums of money on the Middletown lead-silver mine, and Colonel James, a British officer, was still in charge of this mining venture in 1775. The British were primarily interested in silver and so were their workers. Local farmers became suspicious of the odd behavior of one of the workers, who after only a short time had amassed sufficient funds to buy some land and construct a house. It was noticed that he frequently thrust the iron rod, with which he stirred the molten lead, into the ground around the kettle. At night he carefully collected the soil that had become enriched with silver and melted it. When the authorities became aware of his nightly endeavors, they arrested him and charged him with theft, but nothing could be proven, and he was allowed to go free.

On the eve of the War of Independence, Governor Trumbull and the Committee on Safety seized the mine and ore that was ready for shipping. On May 23, 1775, Silas Deane wrote Titus Hosmer: "The state of the lead mine in this town [Middletown] has likewise engaged our attention. Upon inquiry, we find the ore is plenty and reputedly rich, the vein is opened, seven tons of ore now raised and ready for smelting, and any other quantity may be had that may be required . . . there can be no reasonable doubt, if we can succeed in refining, that this mine will abundantly supply not only New England, but all the colonies with lead, in such plenty as to answer every demand of war and peace."

The Connecticut Assembly directed Jabez Hamlin, Matthew Talcott, and Titus Hosmer "to provide stores of lead as they should judge necessary for the use of the Colony." Captain Samuel Russell was appointed superintendent of the Middletown mine, and Governor Trumbull informed General Washington that seven or eight tons of rich ore had been mined and smelting would soon begin. On July 16, 1775, the committee received instructions to give one thousand pounds of lead to Joseph Webb, to be delivered to Captain Jeremiah Wadsworth for the army at Cambridge. In November, the committee reportedly possessed 5,140 pounds of lead, ready for distribution to the towns. The General Assembly ordered all of it cast into bullets. The following year, the committee sent two thousand pounds of lead to Nathaniel Shaw in New London and five thousand pounds to General Ward in Newburg on the North River.

In the summer of 1776, the Council on Safety noticed a sudden inflation in lead prices, probably the result of some profiteering, and fixed the price at six pence per pound.

Insufficient knowledge of the technologies involved in mining and smelting caused major interruptions in the production of lead. In November 1775, Silas Deane attributed problems in the mine to the weather: "Our lead works have been kept back by the great rains this month past," he wrote, referring most likely to groundwater flooding. According to accounts by Dr. Barratt, the mine had been laid out scientifically and was initially worked by "regularly educated" miners, probably of Cornish descent. They were promptly "laid off," however, when the war started, and with them went the experience to run a mine. Because groundwater infiltration does not appear to have caused any serious problems for the original miners, Deane's remark was probably intended to disguise his ignorance of mining processes.

Aside from lead, the revolutionaries also had an urgent need for gunpowder. With regard to the latter, Silas Deane wrote on May 28, 1775: "We have resolved to give a bounty of ten pounds on every fifty weight of saltpeter and five pounds on every 100 pounds of sulfur manufactured from materials found in the Colony." The shortages he betrays here are surprising because the lead came from ore that also contained lots of sulfur that was released during the refining process and could easily have been reclaimed.

The Middletown lead vein was several feet thick and mined to a depth of about one hundred feet. Underground passages ultimately stretched horizontally over a length of about fifteen hundred feet. Mining and smelting continued until February 1778, when a report to the Connecticut Assembly stated that the manufacture of lead at Middletown had become "unprofitable" and "was far from the expectations of the public and the salutary purposes for which it was designed." The Committee on Safety thus ordered the works closed, and the state thereby pulled out of a strategic war industry, apparently for economic reasons alone. Up to that time, the mine had produced some 16,000 pounds of lead, enough to pour 200,000 (0.70 caliber) or 600,000 (0.50 caliber) musket balls. Because battlefields had shifted south and the transportation of heavy lead was difficult, it made sense to reduce production in Connecticut, but closing the state's principal source of bullet lead in the middle of the war appears to have been premature.

When work in the Middletown mine resumed in 1850, it became obvious that frugality and/or ineptitude had caused its closure, not the resource depletion hinted at seventy-two years earlier. In 1853, James Percival, the state geologist, and Professor Josiah Whitney, who was later appointed the state geologist for California, wrote extensive reports on the Middletown lead-silver occurrence. Percival mentioned that some specimens of galena collected in the old mine were unusually large and pure and referred to "a great [single] mass, rich in strings of silver ore, which weighed from 1500 pounds to a ton," lying near the front door of the field office. Infected by the geologists' enthusiasm, investors founded the Mattabassett Silver-Lead Mining Company, which started operation that same year (plate 6, top). Whitney reported that the company employed about thirty-five miners and laborers, who raised, on average, one ton of ore daily, and some forty to fifty tons awaited transport at the time. The depression of 1855 appears to have been the principal reason for this company's bankruptcy after a relatively short period of activity.

New England revolutionaries also relied on a lead deposit in southwestern Massachusetts, near Loudville, on the western margin of the Central (rift) Valley. The Manhan lead vein had been discovered in 1679, but mining started on it in earnest in 1765. Ethan Allen and others took possession of the works after the Declaration of Independence, and lead bullets were cast there during the Revolutionary War, the War of 1812, and the Civil War. The last transfer of ownership occurred in 1865. This mine ultimately reached a depth of about one hundred feet, where the miners experienced serious problems with groundwater infiltration. The owners decided to dig a tunnel eleven hundred feet long in the side of the hill to take care of the problem. The expense of this prodigious undertaking and the competition of cheap lead imported from Missouri led to the mine's closure before the end of the nineteenth century.

A rather unusual Connecticut source supplied yet more revolutionary bullets. Following the successful repeal of the Stamp Act in 1766, grateful New Yorkers honored George III by erecting a gilded lead statue of him on the Bowling Green (now Bowling Green Fence and

Park in lower Manhattan). The king was depicted as the famous Roman emperor Marcus Aurelius on horseback. On the night of July 9, 1776, news of the Declaration of Independence reached New York, and a rowdy crowd gathered on the green. By the light of numerous campfires, the colonists tore George's statue from its pedestal, cut the head off, and triumphantly paraded it on a spike through the streets. The British later "rescued" the head and sent it to London as an example of colonial barbarism. The horse, however, remained with the colonists to be chopped up and transported piecemeal in oxcarts to the Litchfield home of General Oliver Wolcott, who had the fragments buried in his back yard under an apple tree. A few months later, the entire Wolcott family and numerous friends occupied themselves with smelting it, producing 42,088 lead bullets! The three Wolcott children, Laura (fourteen), Mary Ann (twelve), and Frederick (ten) poured and molded about half of this number, significantly contributing to the war effort while their father was away fighting. These bullets were probably used in the crucial Battle at Saratoga, where Bourgoyne's southern advance was stopped in the fall of 1777. His Majesty's statue was in this way returned to His Majesty's troops with the compliments of the rebels.

Emeralds in the Middlesex Hills

Gold in the Cobalt Hills east of the Connecticut River and silver (in lead) along the river's western bank near Middletown raised the inevitable question: Could there be more mineral treasures in that region? Frederick Hall, a professor at Middlebury College in Vermont, knew that the answer was yes. In July 1838, he paid the fifty-cent fare and boarded a steamboat from New York to Hartford. As the boat made its way up the Connecticut River, Hall noticed "the barren hills, alive with human beings hard at work." The sounds of drills and sledgehammers echoed throughout the valley. Piles of stone heaped along the river's edge awaited shipment to New York and markets as far south as Savannah and New Orleans. Here and there, white masses protruded from the dark bedrock in deforested sections of the hills, outcrops of pegmatite veins emplaced along faults that intersected the metamorphic gneiss complexes (fig. 3–3). The veins range in width from a few inches to as much as one hundred feet. Their appearance in the landscape was

Fig. 3–3. Pegmatite vein that intruded along reverse (high-angle thrust) fault in a gneiss complex exposed along route 9 near Essex.

a sign that the boat was approaching Haddam, the southernmost area of the famous Middlesex pegmatite district (fig. 3–4).[2]

Hall begged the captain to put him ashore, saying that he simply could not pass up a visit to Haddam, "a place known all over the civilized earth, for the richness and variety of its minerals." As soon as he landed, Hall hired two quarrymen to help collect (chryso)beryl crystals. A well-known pegmatite vein was naturally exposed near Haddam's meetinghouse that continued right into its cellar. The most promising spot in the vein—with crystals clearly visible at its surface—was located only four feet from the northern side of the meetinghouse. After some haggling, Hall agreed with the building's owner, a

2. The Middlesex pegmatite district extends from Deep River to Glastonbury, spanning a distance of more than thirty miles. In the nineteenth century, more than one hundred quarries were worked in this region for various minerals, especially feldspar and micas. The gneiss provided building blocks for buildings, pavements, etc. A second important pegmatite area, equally rich in rare and gem-quality minerals, is located near Branchville, north of Norwalk.

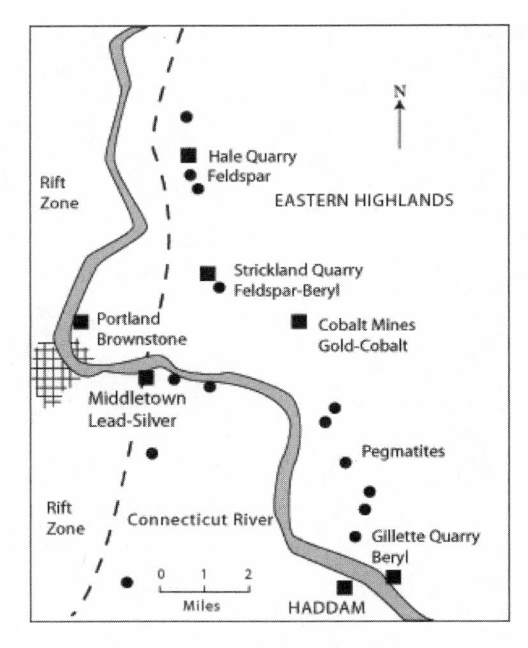

Fig. 3–4. Distribution of principal quarry/mining sites in the Middlesex pegmatite district. Dots represent various pegmatite quarries; squares indicate important sites at which, respectively, beryl, brownstone, feldspar, cobalt/nickel, and lead/silver were extracted.

Mr. Brainard, to pay five dollars for a "single blast" at the spot, assuring the man that he would take care of any damages.

Using hammers and chisels, the workers drilled a hole about a foot deep and loaded it with a pound and a half of powder, and then everyone took cover in a nearby cornfield. Away went the charge and part of the outcrop shattered, with fragments flying in all directions. The explosion actually hurled a quarter-ton rock right over the meetinghouse, depositing it some thirty feet beyond. Aside from a few broken windowpanes, amazingly, there was no damage to the building. Satisfied with the results, Hall collected various minerals, among them green tourmalines and golden crystals of beryl. After paying his dues, he ended his account of collecting beryls in Haddam with: "I have done with Haddam."

Beryl is a beryllium silicate mineral that crystallizes in six-sided columns.[3] Its name derives from the Sanskrit word for *brilliant*, and when it is translucent, beryl produces several precious gemstones. Golden yellow beryl is known as heliodor (plate 7), the blue variety is called aquamarine, and green crystals provide emeralds. There is also a rare pink

3. Beryl is a relatively simple but rare silicate mineral ($Be_3Al_2Si_6O_{18}$). Beryllium, the lightest of all metals, boasts considerable structural strength and a high melting point. Beryllium alloys are widely used in the aerospace and electronic industries.

variety known as morganite. Well-formed crystals usually grow in cavities and can in fact become quite large, but they lose their transparency.

Harold Stearns, who entered Wesleyan in 1917, paid for much of his education by collecting and selling pegmatite minerals. On Sundays, he would take the trolley across the Connecticut River to Portland and spend much of the day at the famous Strickland Quarry (figure 3–4). Mining there concentrated on feldspar and sheet mica, which was used mainly for windows in stove doors. The workers were not interested in rare minerals but often put them aside for rockhounds. Stearns found many beryl crystals, though only a few contained gem-quality material.

Stearns's most valuable find was a four-inch-long transparent tourmaline crystal that was dichroic: held against the light one way, it was blue; rotated ninety degrees, it turned green. Stearns decided to sell it and took the boat to New York. There he met with several dealers, only to learn that similar specimens could be obtained in Brazil for the price of a Cuban cigar. Disappointed, he then went to see George Kunz, a well-known mineralogist at Tiffany's. The enterprising student impressed Kunz, who hired Stearns at twenty-five dollars a week to seek rare minerals in the Middletown pegmatite district. One day, when visiting the Gillette Quarry in East Haddam, Stearns found a flawless golden heliodor weighting 301 carats. He decided to keep the crystal and had it cut into a series of gems for a necklace. Stearns continued to collect and did not "get done with the pegmatites" until his graduation in 1921, after which he became a sought-after geologist in Hawaii.

In the period 1905–9 the American Gem company, an affiliate of Tiffany's of New York, actually leased the Gillette quarry. Later, before World War II, J. P. Morgan, the wealthy financier, became particularly interested in gem-quality tourmalines and continued the lease. Among the numerous gem-quality minerals this site would yield was a slender, 9.5-inch-long tourmaline (elbaite) crystal, triangular in cross-section and rich green in color; it wound up in the American Museum of Natural History in New York. Tourmalines are among the most varied and chemically complex of all minerals and are therefore of great interest to scientists and collectors alike.[4] Most

4. Tourmalines ($[Na, Ca]\ [Mg, Fe^{++}, Fe^{+++}, Al, Li]_3\ Al_6[BO_3]_3\ Si_6O_{18}\ [OH]_4$) are "wastebasket" minerals. When tourmalines are warmed by rubbing they become positively charged at one end and negatively charged at the other, an electrical phenomenon often used to great effect by nineteenth-century magicians.

common are the black tourmalines; what they miss in color is compensated by their perfectly shaped crystals (plate 8). When Morgan's lease expired, Gino Vitali, an avid mineral collector from New Jersey, paid several visits to the quarry. Among numerous crystals, he found a light green transparent beryl measuring two by two inches and a cluster of pink beryls. To Vitali, this outcrop contained the "Queen of all Connecticut pegmatites."

Gems in Quarry Tailings

When studying geology in the old university town of Utrecht in the Netherlands, I spent many afternoons bent over drawers full of minerals, trying to learn their exotic names in preparation for an oral exam. Among the crystals were a small greenish beryl and a black tourmaline that stood out from the others because of their perfect shapes. . Their tags revealed that they had been collected in Haddam, Connecticut. Little did I know that I would live in that village a decade later and roam its ledges and woods in search of similar specimens. Collectable minerals can still be found in several outcrops and the tailings surrounding old quarries, but rockhounds have found almost all of the good specimens, and I never encountered any clear enough to be faceted into gems.

While teaching at Wesleyan University, I frequently took students and schoolchildren to the abandoned Strickland quarry/mine in Portland. One day, after a good rain had cleaned the dust of the minerals in a huge waste pile, a third-grader proudly showed me the treasures she had collected: yellow feldspars, smoky quartz, silver muscovite flakes, reddish-brown garnets, and a somewhat rounded dark green crystal that was about one inch long. Expecting that the crystal was a piece of tumbled bottle glass with which rockhounds liked to seed the area, hoping maliciously to trick fellow collectors, I said, "Nice, great crystals, keep looking," and forgot about it.

Sometimes, however, the mind makes curious circuits. Fifteen minutes or so later, I realized that I had seen a little black fleck inside her "glass." Real crystals can contain inclusions, but glass usually does not. I quickly located the girl on the other side of the waste pile and asked to see her treasures again. Her small dirty hand fished once, twice, three times into deep pockets; at last, the green mineral appeared. I looked at it through a magnifying glass. Surprise! It was a beautiful green beryl crystal, part of it gem-quality. In one half, it contained an intergrown biotite flake, which suggests that the mineral had grown in the contact zone between the pegmatite and a biotite-rich gneiss. Cut and faceted, the other half would make a pretty little emerald. I told her to carefully wrap it in paper and surprise her mother.

I returned many times to this site before it was turned into a golf course and most of the tailings were used to fill a deep mine shaft. I found quite a few beryls but never one with emerald quality.

Among the collections made by John Winthrop Jr. in the lower Connecticut River Valley appeared some black crystals he had found inside a pegmatite that was most likely exposed in the Maromas section of Middletown. He named the crystal columbite, to honor Columbus and its source in America. His grandson sent Winthrop's collection to England, where it lingered in museum drawers until Charles Hatchett, an English chemist, found the sample in 1801 and determined it to be an oxide of an undiscovered element he called columbium. A well-known mineralogist, William Wollaston, then compared the specimen with tantalum-oxide minerals and concluded that Hatchett's columbium was the earlier-named element tantalum. Later still a German chemist, Heinrich Rose, decided that they were indeed different elements. He gave columbium a new name, niobium, which derives from Niobe, the daughter of Tantalus, because the elements named after them were so much alike. In this way a solid American name for a rare element with Connecticut origin was lost.

The most complex mineral among the more than 140 that occur in Middlesex pegmatites is Samarskite, a rare earth oxide named after a Russian Colonel, Vasilii von Samarski. This amazing mineral contains yttrium, cesium, uranium, niobium, tantalum, and titanium. Its velvety black radioactive crystals are opaque and were once fashioned into cabochons, though this is not the best precious stone to carry close to one's heart!

Lithic Resources: Soapstone and Quartz

Southern New England had ice-free summers by around seventeen thousand years ago, but its climate remained harsh, and the rapid melting of the ice sheets and fast changes in fluvial patterns created a very diverse landscape. Hills capped by the polished surface of freshly exposed bedrock alternated with valleys choked with glacial debris. Lakes dotted the landscape, and because of rather chaotic drainage, wetlands were also abundant. This beautiful but barren land, then, patiently awaited the arrival of the earliest life forms.

Between fifteen and fourteen thousand years ago, healthy herb-shrub tundra established itself on the sediments left by the glaciers. With these plants came the first animals, ranging from mastodons and caribou to beavers and hyrax (comparable to alpine rodents). Native American hunters are believed to have entered New England after about eleven thousand years ago. The earliest Paleo-Indian site, uncovered near Washington, Connecticut, in the western highlands, is dated at 10,190 BP (before the present). It was a small campsite used by nomads who most likely followed migrating herds and retreated south in wintertime. The tools those hunters used were made predominantly from chert, chalcedony, and jasper, fine-grained rocks that provided flakes with sharp cutting edges. No such rocks crop out in Connecticut, although glaciers might have dragged some cobbles of chert southeast. Much of the chert was instead most likely collected in New York State, and the jasper in the Boston area, indicating the hunters' high degree of mobility.

The period between 9,000 and 7,000 BP is known as the Early Archaic. Mixed conifer-deciduous forests had migrated north into southern New England by this time, and the environment had become increasingly less hostile to humans. New types of projectile points, some fashioned from local materials, made their entrée. Their appearance could point to an influx of new tribes from the south, as new types of tools suggest an influx of different cultures.

The Middle and Late Archaic periods lasted from about 7,000 to 4,000 BP and included a period of global warming. The higher temperatures increased evaporation and transpiration rates, which caused drier conditions, and the benefits of the warmer climate were somewhat negated by the impact of droughts, especially in the highlands. The topographically controlled differences in the moisture content of the soils led to a more varied fauna as well. Humanity's exploitation of an increasingly diverse array of natural resources resulted in more balanced diets and therefore higher survival rates, and the rapidly increasing populations then began to organize themselves into tribes. Late Archaic cremation burials point to the emergence of common beliefs and an increasing tendency toward cultural identity.

Indians needed lithic materials to produce and replace their tools and weapons. The best rocks for such specialized uses should be hard, fine-grained, strong, and relatively common. The first three properties are characteristic of many silica-rich volcanic rocks, such as rhyolite and obsidian. At one time, these rocks might have been exposed in Connecticut, but with exception of a small occurrence near Trumbull, they

Fig. 3–5. Chert spearhead (left) and quartz arrowheads from site along Salmon River in the Moodus area (private collection).

disappeared through erosion many millions of years ago. As native groups became more sedentary, they increasingly needed to rely on what they could find locally. Most abundant were the rounded cobbles of milky-white quartz that stand out in virtually all New England streams and can be chipped into sharp-edged flakes (fig. 3–5).

Vein quartz like this is the product of hot water circulating underground in fault zones during periods of regional metamorphism and/or magmatism. Water heated at depth dissolves silica from the surrounding rocks. As these fluids rise, they cool and their solubility decreases, allowing tiny quartz crystals to nucleate on fracture surfaces. This process is similar to the crystallization of sugar on the bottom of a cooling cup of tea. As the water continues to circulate underground, there is a more or less continuous supply of silica and oxygen, and the fractures slowly fill with growing crystals. When the ice sheets scraped across such veins, they loosened angular chunks that were then ground down and finally rounded during their transport in streams.

Vein quartz is indeed a hard, dense rock, but it has significant disadvantages for tool making. Although few fractures occur naturally inside single crystals, breaks between the crystals are common. Therefore, fashioning relatively small arrowheads is easy but producing larger fragments, such as those needed for spears, is virtually impossible.

Sometime in the terminal Archaic period (2,500–2,000 BP), the Indians stumbled onto a providential new resource: steatite, commonly known as soapstone. William Fowler, a Massachusetts archaeologist, believes that the extraction and implementation of soapstone occasioned a major cultural upheaval. It represented the first major step from nomadic to more sedentary ways of life; it created new customs; and it stimulated intersocial relations through contact between tribes by trade. For more than one thousand years, this stone treasure was quarried for both common and ceremonial use.

Soapstone, is a soft, silky, fibrous rock that can be used to produce bowls, cups, and ladles as well as sinkers, pipes, and decorative beads. Finds of soapstone bowls and fragments on Cape Cod, far removed from original quarry sites in Massachusetts, suggest long-distance trade patterns and indicate their great value. Several cremation sites have yielded the remains of soapstone bowls that have been "killed," or broken on purpose, suggesting that they were not purely utilitarian but also highly symbolic, especially in religious ceremonies. One reason for their special status might have been that soapstone outcrops are rare and the production of soapstone bowls is difficult and time consuming.

The use of soapstone in food preparation represents a similarly major breakthrough in "kitchen" technology. Previously, food that was not prepared over an open fire was usually "stone-boiled" in vessels made from birch bark, elm bark, or animal skins: the vessels were filled with water, ground nuts, and/or other ingredients, and heated stones were added to the mixture. This method was undoubtedly refined over thousands of years but remained primitive and must have led to frequent accidental spills and burns.

During the winters, when people depended more on woodland resources such as acorn, beech hickory, and chestnut, it became necessary to prepare gruels and keep food warmer for longer periods. Soapstone bowls would have satisfied this need. They are sturdy, almost unbreakable, heat retentive, and able to withstand direct exposure to hot ash beds. It is not uncommon for these bowls to retain heat for more than a day. However, not every wigwam had a bowl! The total number of bowls and fragments that have been dug up is relatively small. This could be due to the difficulty of finding native sites, but more likely the bowls were status symbols used primarily during important ceremonies, both festive and funereal. They may have added to but would not have initiated a sedentary lifestyle.

For one thing, as mentioned previously, soapstone outcrops are rare. In Connecticut they occur mainly as small, elliptical masses arranged in two curvilinear belts like strings of beads. The western belt, which contains most outcrops, extends from the Connecticut coast north through western Massachusetts and into Vermont and then Quebec. There might not have been more than twenty or thirty Indian quarry sites altogether in southern New England. The largest concentration in Connecticut is in and near the Tunxis Valley. Despite much destruction from later quarrying activities, several outcrops still reveal traces of Indian work. At one site near Westfield (Massachusetts), archaeologists found a complete cache of Indian quarry tools, including quartzite hammers, chisels, and files.

To make a large bowl (some weighed more than fifty pounds), the natives started by digging an oval ditch, one to two feet in diameter, into the glaciated surface of an outcrop or glacial boulder (fig. 3–6). At a depth of four to six inches, they scraped the inner wall of the ditch until they had obtained a shape resembling that of a giant mushroom. When the stem was sufficiently narrow, it was broken through, and the "mushroom" cap turned upside down. In this position, the interior was scraped and chiseled out until a bowl emerged.

William Fowler, an archaeologist, described the worked soapstone boulders east of the Connecticut River near Wilbraham, Massachusetts: "It is apparent the Indians hewed out the solid bowl-shaped blanks from the parent boulder, and carried them to the top of the knoll, where the rougher part of the hollowing process was carried on. Apparently no long period of occupancy occurred, for only a few arrow points and fire-cracked stones were found, suggesting a few days' camp while the work in the quarry was in progress." Obviously Fowler had never himself attempted to excavate a soapstone bowl with quartzite tools; though soapstone is soft, the chaotic interfingering of its tiny, needle-shaped crystals makes it strong and tenacious. The excavation and preparation of a single bowl in fact would have taken several months.

Fig. 3–6. Top: soapstone (steatite) bowl. Mashantucket-Pequot Indian Museum. Bottom: Outcrop of soapstone worked by Indians. Note the oval/rounded ditch carved around the bowl with quartzite tools and broken central stem.

Soapstone usually occurs in association with serpentinites, dunites, and amphibolites. The latter rock types are commonly referred to as ultramafics, because they are relatively low in silica and relatively high in iron and magnesium, which in turn cause their dark colors. Ultramafics are believed to reside in the Earth's uppermost mantle and lower crust. When plates collide, a mosaic of crustal blocks, or terranes, separated by major fault zones, or sutures, develops in the collision zone. Some of these faults break through the brittle crust and tap into the more ductile layers at depth. Blobs of ultramafic material become mobilized and are squeezed up along these suture zones in the form of a crystal mush, composed mainly of olivines and pyroxenes.[5] Because olivines are heavier than pyroxenes, the crystals may separate as they rise, ultimately forming rather distinct masses.

As these masses continue their rise, they react with water and carbonate gases from adjacent rock formations, and new minerals form. Olivine is transformed into serpentine; pyroxenes change into amphiboles.

5. Olivine ($[Mg,Fe]_2\ SiO_4$) and pyroxene ($[Mg,Fe]\ SiO_3$) are silicate minerals commonly transformed into serpentines and amphiboles at shallow crustal levels.

The incorporation of water into the crystal lattices of these new minerals leads to a significant volumetric increase in the intruding mass, which provides a further impulse for the upward movement of the crystal mush inside the fault zone. Repeated shearing of the serpentinized masses during slippage in fault zones can lead to the localized formation of steatite (soapstone) and, ultimately, pure talcum.

Serpentine crystals take many forms: some are platy (antigorite); others are fibrous and consist of tiny needles (chrysotile). The latter variety is also known as asbestos. In Rome, where this material was used for wicks in oil lamps, it was referred to as "amiantus," or pure, incorruptible matter. According to the Roman poet Orpheus, serpentine was also believed to have medicinal properties:

> Let him who by the dragon's fang has bled,
> On the dire wound serpentine powdered spread,
> And in the stone his sure reliance place,
> For wounds inflicted by the reptile race.

New England's eighteenth-century farmers frequently used rectangular slabs of soapstone, preheated in the ashes of their fireplaces, to warm their beds, while remaining unaware of the rock's industrial potential despite its obvious heat-retention properties. This is evident from Professor Frederick Hall's recollection, early in the nineteenth century, of the words of a bystander as he collected a few specimens of serpentinite along a road west of Westfield, Massachusetts. The man, amazed at the professor's interest in the rocks, grumbled, "That hill of rocks there is good for nothing, and ever will be—a worthless waste—a mere dead weight placed there Saturday night, after Creation was over, to be forever in man's way."

A few decades later, this site became part of a large complex of quarries, and serpentinites staged a major comeback. During the Industrial Revolution, the use of asbestos expanded exponentially because of its low heat conductivity, high resistance to electricity, and chemical inertness. Major deposits were discovered in Vermont, and by the mid 1940s its mines and quarries produced more than 90 percent of the asbestos needed in the United States. As early as 1903, articles discussing the role of asbestos dust in causing mesothelioma, a lung disease, appeared in such journals as *Scientific American.* More than half a century of legal maneuvering ensued, and thousands of people died, before laws protecting both workers and the public were enacted. A millennium ago, some native women may thus have suffered from the same medical conditions that became endemic among workers in the asbestos quarries and factories of the Industrial Age.

The discovery of soapstone pods was as important to the natives as the finds of lead veins were to the colonists. They initially caused minor changes in the lifestyles of small groups, but this was followed by a ripple effect: a regional demand and usage of those resources. Soapstone bowls made it to Cape Cod and Buffalo; lead bullets to Boston and Saratoga. Both were catalysts for important historic events.

Other Historic Quarries and Mines in Connecticut

Stories about interesting metal and mineral resources in this chapter have come primarily from the Middlesex District in south-central Connecticut. As shown in figure 1-2, several other mining/quarrying districts and sites with much greater historic impact than those discussed so far developed in Connecticut. Most important among those were the Salisbury Iron District, the Old Newgate Copper mine near Granby, the Portland Brownstone quarries, and the Stony Creek Granite District of Branford and Guilford. Their histories have appeared in several excellent publications. To include abstracted accounts of these works in *Stories in Stone* would have done a disservice. The following is a short list of the books and articles that I consider to be of special interest and significance:

On Iron: R. B. Gordon, *A Landscape Transformed: The Iron Making District of Salisbury, Connecticut* (2001); K.T. Howell and E.W. Carlson, *Men of Iron: Forbes and Adam* (1980); E.W. Carlson, *Empire over the Dam* (1974); H. C. Keith and C. R. Harte, *The Early Iron Industries of Connecticut* (1935).

On Copper: C. R. Harte, *Connecticut's Iron and Copper* (1944); R. Phelps, *Newgate of Connecticut: Its Origin and History* (1895).

On Brownstone: A. C. Guinness, *The Portland Brownstone Quarries* (1987).

On Granite: D. Deford (ed.), *Flesh and Stone: Stony Creek and the Age of Granite* (2000).

On Mining in General: J. A. Pawloski, *Images of America: Connecticut Mining* (2006); G. L. Studley, *Connecticut the Industrial Incubator* (1982).

On the Geology of Connecticut: J. W. Skehan, *Roadsite Geology of Connecticut and Rhode Island* (2008).

Settlers and Soils in the Central Valley

The Legacy of Glacial Lake Hitchcock

Nothing can exceed the Beauty, and Fertility of the Country. The Lands upon the [Connecticut] River, the flatt low Lands, are loaded with rich, noble Crops of Grass, and Grain, and Corn.
—John Adams, 1771

The geology of the Central Valley in western Massachusetts and central Connecticut profoundly influenced native and colonial settlement patterns in southwestern New England. In the Valley, the land is generally flat and arable, soils are relatively rich, and access to the river is easy. Save for the area's highlands, where farmers were plagued by glacial debris littered with stones, settlers in the Valley were able to progress rapidly from a subsistence phase into horticultural and agricultural societies. In 1906, historian George Roberts wrote, "When Nature produced New England she was a philanthropist for she was bountiful in her beneficence. When she produced the valley through which the Connecticut flows for three hundred and fifty miles, she was an artist—*the artist.*"

Almost eighty years later, Michael Bell was less poetic and more specific. In *The Face of Connecticut,* he wrote, "Had there been no rift valley to store the sediment, there would be no brownstone. Without brownstone, the Central Valley would have been just as resistant to erosion as the rest of the State and would not be a valley. Had there not been a valley, glacial Lake Hitchcock would not have formed here. And without the lake deposits, there would have been no fertile farm land to draw the early settlers like a magnet." Bell referred to the Central Valley as the great "incubator" of Connecticut.

Two major geologic events made the Central Valley what it is: the development of a tectonic rift valley during the late Triassic/early Jurassic periods, and the arrivals/retreats of massive glaciers during the Pleistocene ice ages.

The Early Mesozoic Central Valley

The earliest geologic processes in the Central Valley date back some 235 million years, when tectonic forces began to pull the North American and African plates apart. Before that time, those plates formed part of a single huge land mass called Pangaea. During the embryonic phase in the opening of the Atlantic basin, the crustal segment between the plates stretched and fractured along old zones of weakness in the Appalachian fold belt. The subsidence of an elongate crustal sliver between two or more major fault zones led to the formation of a rift valley, a long, relatively narrow basin, encompassing more than a thousand square miles and extending from New Haven, Connecticut, to Northfield, Massachusetts (fig. 4–1).

Erosion of the adjacent highlands in the subtropical climate of the era washed large volumes of sand and clay into that depression. This deposition generally kept up with the slow subsidence of the basin floor, until many thousands of feet of sediment had accumulated. About 200 million years ago, voluminous masses of magma rose along new faults that had tapped reservoirs at depth; on three occasions, lava flows filled the rift valley. More sediment was deposited following each of these volcanic episodes. Together, the brownstone formations and lava flows ultimately reached a thickness of more than twelve thousand feet and resembled a giant ribbon cake.

About 170 million years ago, active rifting stopped inexplicably (had it continued, eastern Connecticut would have become part of the African continent). Spreading and crustal thinning did continue along the eastern margin of present-day North America, and the Atlantic basin then slowly opened. Following a period of repose, the contents of Connecticut's rift valley tilted eastward, exposing the older sediments (red

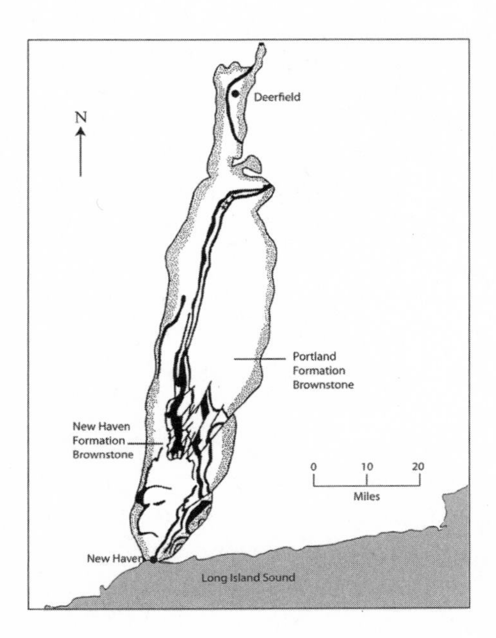

Fig. 4–1. The Rift valley that developed during the early Mesozoic in central Connecticut and western Massachusetts. Its contents were tilted east, which exposed the oldest, late Triassic brownstone along its western margin, and youngest, early Jurassic sediments along its eastern margin.

beds) along the western border, the volcanic section (lava flows and interlayered sediments) mainly in its axial part, and the youngest sediments (brownstone) along the eastern margin.

The relatively soft sandstone formations eroded more easily than the three lava flows and the protected sediments that sat between them. As a result, long, narrow, secondary valleys floored with sandstone developed along the western and eastern margins of the rift valley (fig. 4–1).

The Remains of the Last Ice Age

About 120,000 years ago, after a relatively warm period, sheets of land ice many thousands of feet thick again began to accumulate in northeastern Canada and slowly crept south across New England. The volume of land ice waxed and waned over time, leading to periods characterized by glacial advances and retreats. During the most recent advance, the ice sheets bulldozed across Connecticut, scraping soils, grinding soft sediment, and polishing resistant bedrock. The metamorphic rock formations in the highlands withstood the glacial onslaught reasonably well, as did the much younger lava flows, but the parts of the Central Valley floored by soft brownstone formations were further excavated and deepened.

The time at which the ice sheet started to retreat is still under discussion. Dating of organic materials suggests that extensive melting of Long Island's terminal moraine began between twenty-four and twenty-three thousand years ago. New dating methods that measure the time since ice-covered boulders became exposed to sunlight provides a younger age. Despite these differences, it can be safely assumed that the deglaciation of Connecticut itself began around twenty thousand years ago. The initial retreat was most likely a slow process that accelerated as the ice sheet thinned.

On average, the ice front pulled back to the north at about 150 to 250 feet per year. Brief advances interrupted the recession, but by about 16,000 BP, southern New England was essentially ice free. Two thousand years later the ice sheets had left the Canadian provinces as well. During melting, the soils and rocks that the ice had picked up on its way south were left behind, coating much of the shaved and polished bedrock with till—a heterogeneous mixture of clays, sands, gravel, and boulders. Those unsorted deposits took the place of the deep, fertile soils that had accumulated over the many millions of years before glaciation. It was not the best possible deal for Connecticut.

During the retreat of the ice, large volumes of melt water washed across the bare lands, changing drainage patterns and rapidly filling natural depressions. Glacial lakes developed in much of Connecticut's territory, the largest of them in the Long Island Sound area behind the terminal moraine of Long Island itself and in the Central Valley of Connecticut and western Massachusetts. The latter lake is named after Edward Hitchcock, a professor at and later president of Amherst College. Lake Hitchcock began to form around 18,500 years ago behind a mile-wide natural dam of glacial debris east of Rocky Hill (fig. 4–2). At the site, an east-west trending ridge of basalt narrowed the eastern side of the Central Valley, and large volumes of sand and gravel carried by glacial streams filled the remaining gap. Behind this natural barrier, Lake Hitchcock grew northward, following the retreating ice margin to reach a length of about 180 miles, stretching from Rocky Hill to Lyme, New Hampshire. The southern half of the lake developed inside two troughs, separated by the Metacomet Ridge and floored by Mesozoic sandstones.

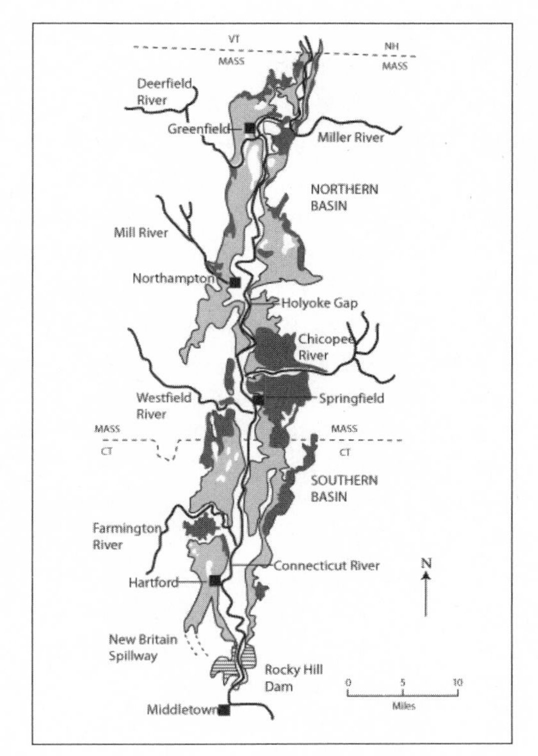

Fig. 4–2. Surficial geology map of glacial Lake Hitchcock. Dark grey areas: major deltaic fans (mainly sands) that formed where rivers entered the lake. Light grey: areas where varved clays accumulated on the lake bottom (*adapted from Stone and Ashley, 1992, figure 1*).

It extended from Rocky Hill to Holyoke, Massachusetts, and had an average width of about ten miles. A spillway in the area of fractured basalt near New Britain controlled the lake's water level.

The northern half of the lake stretched from Northampton into New Hampshire and Vermont. Its widest part, about twelve miles from east to west, was in the Hadley-Northampton area, where it connected with the southern basin via the Holyoke gap, a fractured area in the Metacomet Ridge (fig. 4–2).

Lake Hitchcock must have been an impressive sight, coated in snow and ice in the winter, filled with blue icebergs that majestically floated south in the spring, fringed by shiny, polished bedrock in the summer, and covered with white-capped waves in the fall when the nor'easters started blowing. At midyear, turbulent streams and rivers that originated in the highlands carried large volumes of milky meltwater loaded with clays and sands toward the lake. Much of the sand accumulated in deltas along the lake's margins, but tiny floating particles—mainly clays—remained in suspension and moved farther from the shores. These clay particles slowly settled to the bottom when the lake froze during the winters and the turbulence ceased.

These seasonal changes resulted in two thin, but very distinct, layers of clay and silt with different colors, grain sizes, and thicknesses. Each couplet of layers is called a "varve" (fig. 4–3); like tree rings, each varve thus represents a full year. The varves in a clay pit near Charter Oak Park in Hartford provided a 454-year record of deposition. Varves taken from a clay pit farther north, near the Podunk River in South Windsor, overlapped the Hartford section by 281 years while adding 417 years to the record with its lower levels. A clay deposit near Chicopee, Massachusetts, partially fit the South Windsor record, beginning with its 789th year, and added a section extending to the 1,120th year. Clay pits farther north provided additional data.[1] By correlating various sections, Ernst Antevs, the first glacial geologist to study varves in North America, was able to estimate the lifespan of Lake Hitchcock at 4,100 years. Recent studies suggest that the lake waters may have covered much of the Central Valley for as long as 3,000 years. North of the Holyoke gap, the lake may have existed an additional one thousand years.

About 16,000 years ago the Rocky Hill dam was breached, and the lake emptied. Some scientists believe that earthquakes generated by the beginning of glacial rebound might have triggered a catastrophic event (fig. 4–4). Others favor slow erosion of the dam. Small lakes and wetlands remained in depressions until a regional drainage system was established and the Connecticut River began to carve its new course through the lake deposits. Exposed lake bottoms remained barren and windswept for an extended period. Windblown dust accumulated in dunes and elsewhere formed loess deposits that provided good footing for pioneering plants.

Fossil remains of mosses, herbs, and shrubs indicate that tundra-type vegetation was well established by 17,500 BP. The fossil assemblage included nitrogen-fixing alpine plants such as *Dryas*, which enriched the developing soils. By 11,000 BP deciduous trees had migrated north and more mature soils began to develop, providing the base for

1. Varved clays were ideally suited for fabricating bricks. Around the turn of the twentieth century, more than thirty clay pits operated in the Central Valley, producing about 300 million bricks annually.

Fig. 4–3. Top: Detail of glacial lake varves, showing succession of annual summer and winter deposits. *Courtesy J. R. Stone, United States Geological Survey.* Bottom: Outcrop of glacial lake varves, East Windsor area. *Courtesy J. R. Stone, United States Geological Survey.*

future generations of native and colonial farmers. The thick blanket of lake clays and the sands of the river deltas along the lake's margins covered the older stony tills and bare bedrock in much of the Central Valley, leaving broad, flat lands devoid of rocks and boulders—a big advantage to farmers who needed to plow their fields.

During the last glacial advance, the weight of the ice sheet, more than two hundred tons per square foot, had depressed New England's

Fig. 4–4. Le Deluge: What the Rocky Hill dam may have looked like when the icy waters of Lake Hitchcock broke through (Figuier, 1864, figure 316).

crust. The magnitude of the down-warp depended on the accumulated thickness of the ice and increased as one moved northward. After the ice melted, the crusts slowly rebounded. Because more ice had accumulated in the north than in the south, the land was differentially uplifted in increasing amounts of 4.7 feet per mile to the north. As a result, the flat tops of the glacial river deltas that formed along the margins of Lake Hitchcock, which originally were at the same lake level, presently occupy different elevations. The lake's original spillway near New Britain is presently 82 feet above sea level. The top of the glacial Farmington River delta is at 130 feet. The delta of the Miller River near Turners Falls, Massachusetts, is at 330 feet (see fig. 2–5). At present, ascending stairs of terraces composed of lake and deltaic sediments border the meandering river and its flood plains. As the land slowly rose and tilted, the new Connecticut River cut deep into the former lakefloor deposits, and new regional drainage systems developed.

The Central Valley has an additional advantage for settlers: its own microclimate, which provides a longer growing season than that of the highlands east and west of this topographic depression. The average duration of a freeze-free season in the valley varies from about 170 to 180 days per year. In the northwestern and northeastern highlands, the season averages about 155 days per year, or two to three weeks less (fig. 4–5, bottom). The optimum period for growing corn in Connecticut varies from 150 to 180 days, and there is not much margin for error in the highlands. Weather changes can easily damage or destroy crops altogether. With the exception of some farms in highland valleys, agricultural yields in those regions were usually much smaller per acre than in the Central Valley.

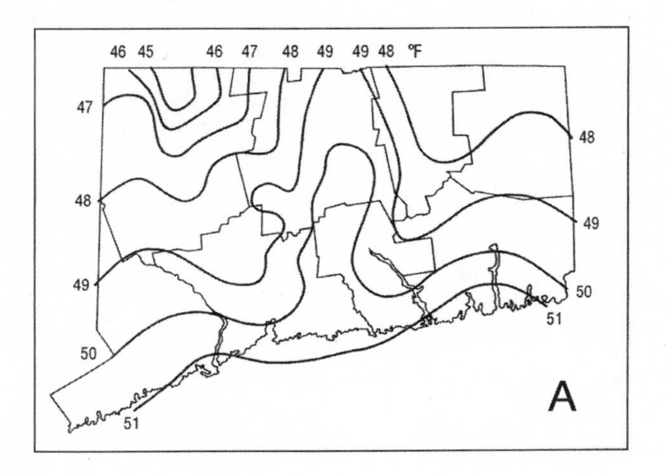

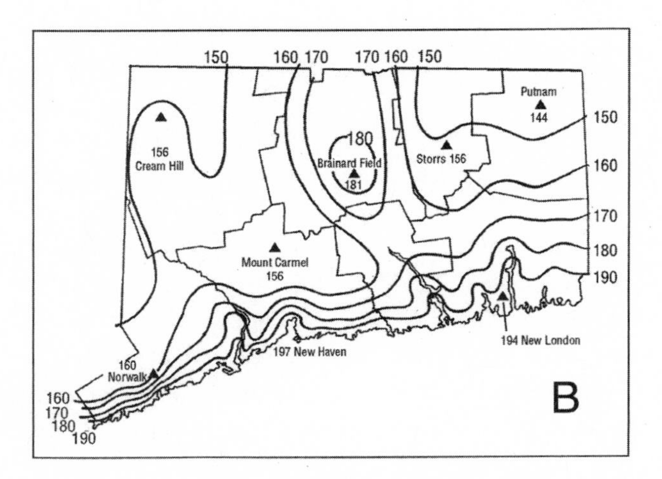

Fig. 4–5. Top: Annual mean temperatures. Bottom: Average lengths of freeze-free season days in Connecticut. Note influence of Central Valley on climate (Brumbach, 1965, figures 7 and 8). *Courtesy Connecticut Geological Survey.*

Indian and Colonial Settlements

The anthropologist Kathleen Bragdon distinguished three southern New England ecosystems that played a dominant role in its cultural history: the estuarine, the riverine, and the uplands zones. Although considered riverine overall, the Connecticut River is actually a 360-mile-long ribbon that intersects all three systems. It has a wide estuary fringed by wetlands, and its lowest segment in southern Connecticut and upper segment in New Hampshire and Vermont are narrow and fringed by steep highlands. Only its middle segment, which stretches from Middletown to Northfield, Massachusetts, is truly riverine. In this segment, the river meanders slowly through wide flood plains that developed on and in the bottomlands left by Lake Hitchcock. The only exception occurs in southwestern Massachusetts, where over a short interval the river was forced to cut through the steep slopes of the trap rock Metacomet Ridge.

In the early seventeenth century, Johannes de Laet reported that although few natives lived near the mouth of the Versche Rivier, they became numerous beyond the narrows near present-day Middletown. There, natives whom de Laet called Seguins grew maize and inhabited villages on both river banks. One large settlement was surrounded by palisades and resembled a fort. Figure 4–6 shows the concentration of Late Archaic (about 5,000 BP) native sites discovered in Connecticut. It is clear that many were located in the riverine section of the Central Valley and at the confluences of the Connecticut River and its tributaries. Reasons for the concentration at the mouth of the Salmon River will be discussed in chapter 6.

Many centuries before the British set foot in Massachusetts, the bottomlands of former Lake Hitchcock and the floodplains of the Connecticut River provided native people with bountiful crops of maize, beans, and squash, the "three sisters" of the native diet. Maize and squash entered the native food supply about nine hundred years ago, and beans entered a few hundred years later. Seven hundred years ago relatively large villages with a horticultural base dotted the riverine section of the valley between Middletown and Deerfield. Each household needed one to two acres of cleared land to grow produce,

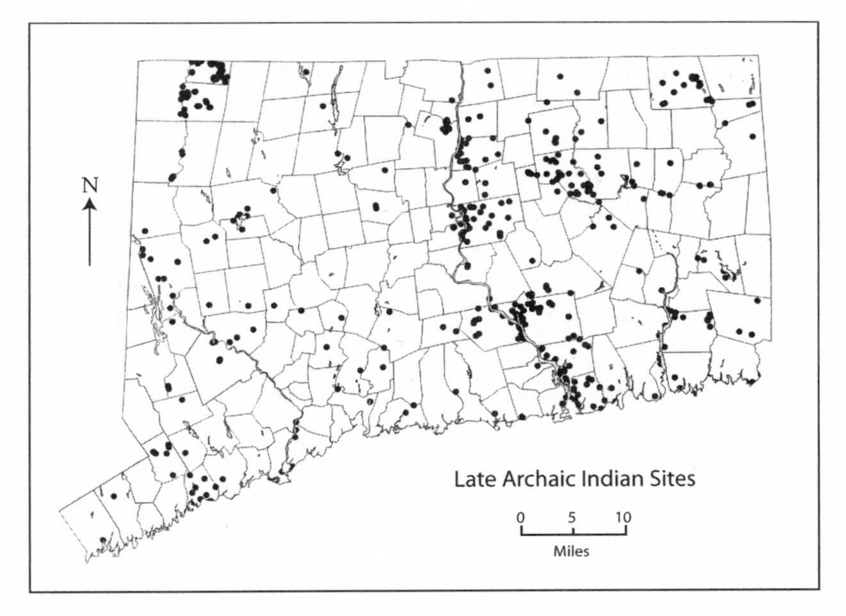

Fig. 4–6. Late Archaic Indian sites. Note concentration along Connecticut and Salmon Rivers. *Courtesy W. E. Keegan, Heritage Consultants.*

suggesting that long strings of plantations would have sprawled along the riverbanks.[2]

The extent to which the natives had cleared the land on either side of the river is reflected by the names early settlers used to describe their new habitations: from north to south, Northfield, Greenfield, Deerfield, Hatfield, Westfield, Springfield, Longmeadow, Suffield, Enfield, Bloomfield, and Wethersfield. Near Windsor, in fact, stood a native village called Poquonok or Pyquag, which translates to "cleared fields."

With the exception of the Dutch, who set up trading posts, few Europeans made incursions into the valley until Adrian Block sailed up the Connecticut River in 1614. Two decades later, this changed rapidly as population pressures and religious discord in the colonies of eastern Massachusetts led to a westward diaspora. Later arrivals in Massachusetts were finding only small lots of marginal land available

2. William Cronon, author of *Changes in the Land,* estimates that as many as 100,000 natives might have inhabited New England around 1600. Colonial populations would not reach a similar number until a century later.

for settlement; the Promised Land had become crowded. The governor of the Plymouth Colony, William Bradford, wrote that the early pilgrims in the Plymouth Bay area were trying to increase cattle stocks to sell to newcomers, and this pursuit had occupied more and more of the best land. He deplored the way the colony was evolving: "No man now thought that he could live except he had cattle and a great deal of ground to keep them . . . by which means they were scattered all over the Bay . . . the town in which they lived compactly till now was left very thin and in a short time almost desolate." Bradford estimated that by 1634 there were fifteen hundred head of cattle, four thousand goats, and "numerous" swine in the Massachusetts Bay area.

New people were thus forced to work even the rockiest fields on eastern Massachusetts hillsides because good lowlands had become scarce. In addition, many settlers had been city bred and were inept at farming and unfamiliar with the food plants that grew in the new territories. They had difficulty adapting to their new way of life. Indicative of such despair is the 1631 letter John Pond sent to his parents in the Old World. "I pray you remember me as your child . . . we do not know how long we may subsist, for we cannot live here without provisions from old eingland."

Having heard glowing stories about wide-open expanses of fertile land from some natives and from Edward Winslow, who had explored the lower segment of the Connecticut River in 1632, a number of colonists requested permission to move west. The Plymouth leaders initially argued that such a push would dangerously weaken their colony, "there being three or four thousand warlike Indians." A smallpox epidemic in 1633 and 1634 changed this situation, however, by wiping out much of the native population in the lower and part of the middle segments of the Connecticut River Valley.[3] Governor Bradford reported that all of the native inhabitants of an unnamed village north of the Dutch trading post Huys de Goede Hoop (present-day Hartford) had died miserably: "This spring, those [natives] that lived about the trading house at Windsor fell sick of ye small poxe, and dyed most miserably; for a sorer disease cannot befall them . . . many of them did rott above ground for want of burial."

3. An earlier epidemic occurred between 1616 and 1619, when a major plague, either chicken pox or viral hepatitis, entered southern New England. It appears to have affected coastal tribes the most.

To the Puritans, the epidemics manifested God's providence. John Winthrop wrote, "God hath hereby cleared our title to this place." Cotton Mather was equally direct: "The New Englanders are a people of God settled in those [lands] which were once the Devil's Territories." By sweeping away multitudes of natives, God made room for the believers.

When news of abandoned villages and treeless stretches of land rich in alluvial soils reached the Bay Colony, a rapid penetration into the Central Valley followed. In 1633, William Holmes led a group from Plymouth and settled near present-day Windsor; a year later John Oldham and a handful of followers choose to make their homes at Wethersfield.

In June 1634, Thomas Hooker, a Newtown (present-day Cambridge) minister, sent six of his followers west on a reconnaissance mission. That same year, he formally requested permission to leave the colony. Town leaders refused, despite the fact that the settlement had begun to suffer for lack of food and other necessities. Two years later, Hooker and about one hundred members of his congregation defied their leaders and left anyway. The official reason for departure, they wrote, was that Newtown had become "toe steight [cramped]" for so great a number of immigrants. According to the historian Jonathan Trumbull, the group made its way more than one hundred miles through a "hideous and trackless wilderness . . . over mountains, through swamps, thickets, and rivers, which were not passable but with great difficulty." Hooker's group herded 130 head of cattle, however, and his wife had to be carried on a litter because of her failing health, suggesting that the settlers actually followed well-established Indian trails.[4] The biggest hardship turned out to be the construction of simple shelters, before winter set in. Broad axes and a few saws were the only tools available.

Thomas Hooker and William Pynchon, the founders of Hartford and Springfield, respectively, chose to settle on terraces left by the former Lake Hitchcock, adjacent to the Connecticut River. They favored the riverine section of the valley not only because of the availability of large areas of cleared land but also because the soils appeared well

4. The story of Hartford's earliest settlers became renowned. In commemoration, Frederick Church painted his masterwork *Hooker and Company Journeying through the Wilderness.* It shows the forested terrain of the eastern highlands in the background and the "promised land" of the Central Valley, flooded by the golden rays of the setting sun, in the foreground.

suited to agricultural development of the type practiced in England. In addition, the river teemed with fish and served as a conduit for necessary supplies from the motherland. Last, but not least, the new colonies were far enough removed from their larger and more powerful neighbors in Massachusetts and Manhattan so as to avoid interference with their religious and political beliefs.

In time the Dutch settlers of Nieu Nederland (also known as Nova Belgica), who according to Trumbull "were always intruders that had no right to any part of the country," were encircled, harassed, and gradually forced to surrender their lands. The Council of the Dutch trading post Huys de Goede Hoop composed the following hopeless resolution in 1642: "Whereas our territory on the Versche Rivier of Nieu Nederland, which we purchased, paid for and took possession of long before any Christians were on said river, has now for some years past been forcibly usurped by some Englishmen and given the name of Hartford, notwithstanding we duly protested, and whereas they moreover treat our people most barbarously, beating them, even to the shedding of blood . . . we have chosen patiently to suffer violence and to prove that we are better Christians."

The shedding of blood refers to Dutch soldiers who were "cudgeled" by English settlers when they protested the destruction of their cornfields around the fort. According to Walter Hard, the English believed that "possession was nine points of the law and that to have was to hold." In 1650, the Dutch gave up and a new border was established between Dutch and English interests some fifty miles west of the Connecticut River.[5] Within thirty years, the four initial English settlements had mushroomed into more than forty villages spread along the Atlantic coast of Connecticut and Rhode Island and into the riverine section of the Central Valley (fig. 4–7).

Until the immigrants could grow imported grains successfully and in sufficient quantities, local horticultural produce remained crucial for their survival. The native peoples planted maize in small mounds spaced about three feet apart. After the corn had sprouted, beans were

5. Attitudes toward the Dutch "settlers" have changed considerably in the last few centuries. Nowadays Dutch flags with their red-white-blue stripes flutter along many roadsides in a clear sign of welcome. Somewhat puzzling, however, are the words OPEN and ANTIQUES printed boldly inside the white band.

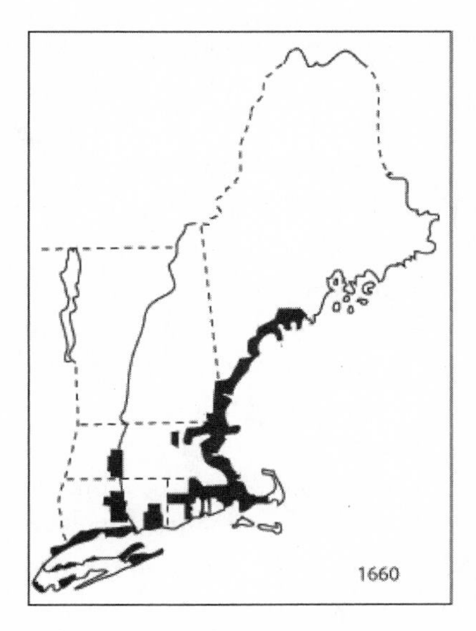

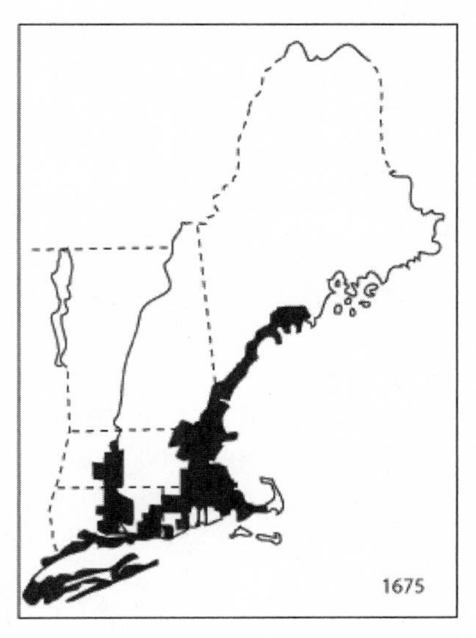

Fig. 4–7. Population concentrations in 1660 and 1675, respectively, showing the importance of the Central Valley during initial settlement of southern New England.

planted around the maize, so their runners could use the corn stalks to climb. Squash then filled in the spaces between the mounds, discouraging the growth of weeds. Reuse of the land for corn, pumpkins, and squash required little nitrogen, and the remains of bean runners provided plenty. The thin coating of wood ash left after the natives burnt the plant remains formed a good base for the next maize crop. When necessary, the natives fertilized their plots with the remains of fish caught in the river and streams during spring runs.

The Morgan archaeological site on the floodplain east of Rocky Hill provides a good example of a riverine horticultural village. Occupation levels range in age from about 1170 to 1360 CE and occur to depths of five feet. Annual spring flooding and its rapid silting protected successive cultural levels from erosion and provided a good stratigraphy, and archaeologists have recovered thousands of stone artifacts here. Many are made of chert, suggesting trade with tribes in the Hudson Valley, where that type of rock is plentiful. By far the most abundant artifacts were fragments of clay pots made from glacial lake clays exposed in Goff Brook, a small stream just west of the site. Most indicative of a successful agricultural enterprise, of course, are the numerous storage pits, several of which still contained the remains of maize kernels. The stratigraphy here points to a multi-seasonal occupation by a farming community over a period of at least two centuries.

Valley native peoples grew significant surpluses too, as an event that followed the outbreak of the Pequot hostilities in 1637 shows. Because native incursions had interrupted farming in the riverine section of the valley, famine threatened the colonial population. The Connecticut General Court hired William Pynchon of Springfield to buy and deliver five hundred bushels of corn to Windsor and Hartford, setting the price at five shillings per bushel. The only river natives that the smallpox epidemic of 1633 and 1634 had not significantly affected were the Pocumtucks, who resided along the banks of the Connecticut and Deerfield Rivers near their confluence. Pynchon contacted the Pocumtucks, but he was either rebuffed or had decided to teach the General Court a lesson in economic reality. Somehow, he insinuated, the natives may have learned that they could use the shortages downriver to negotiate a better price. When Pynchon informed the General Court about his "problems," the authorities dispatched Captain John Mason, who only a year earlier had led a massacre of the Pequots near Mystic. Mason met with a delegation of Pocumtucks in March 1638 but could not induce them to accept the set price and returned to Hartford, convinced that Pynchon had influenced the Indians. The court agreed with Mason's assessment, charged Pynchon with "unfaithful dealing in the trade of corne," and fined him forty bushels of corn.

In May 1638, Mason went north again, this time accompanied by a squad of soldiers. That expedition failed as well, most likely because he had underestimated the number of warriors the Pocumtucks could bring to the field. The General Court ultimately agreed to pay twelve shillings a bushel. Soon, native women carrying baskets brimming with corn filed down the narrow footpaths to the river and filled a fleet of canoes.

The fact that Pocumtuck tribal members could sell such surpluses in late spring indicates an advanced and successful system of horticulture. Five hundred bushels of corn represents a crop of twenty to thirty acres of tillage. Such agrarian abundance was possible only on the fertile soils of the former Lake Hitchcock's bottomlands and the floodplains of the Connecticut River.

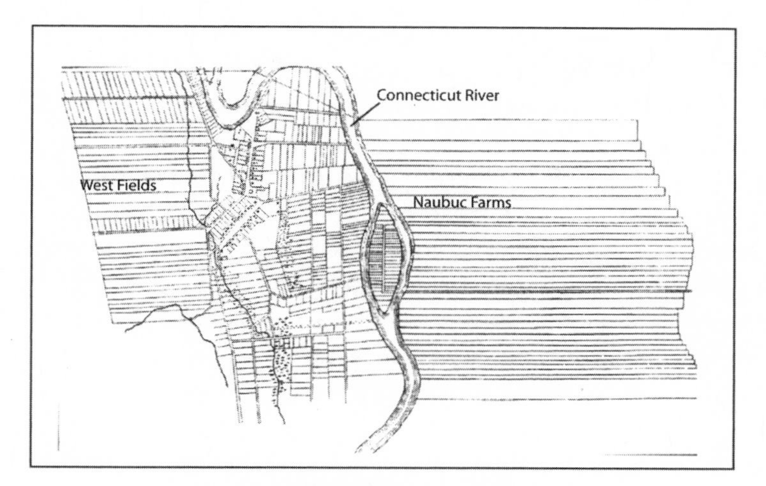

Wethersfield Allotments 1640-1641

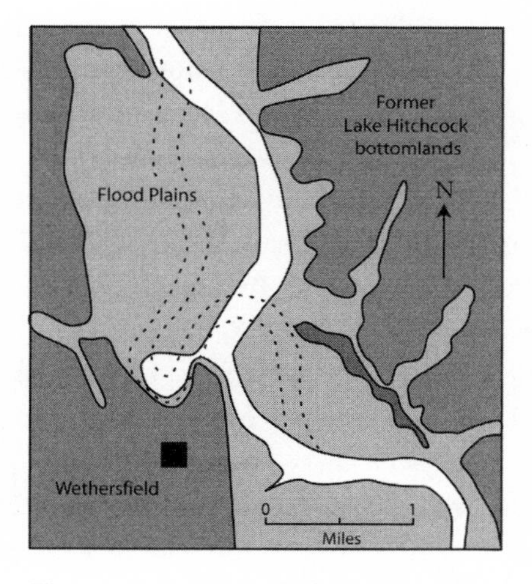

Fig. 4–8. Top: Wethersfield (settled 1634) showing allotments, 1640–1641 *(Andrews, 1889, p. 331)*. Bottom: Approximate position of Connecticut River channel in the seventeenth century and the extent of floodplains and Lake Hitchcock bottomlands near Wethersfield (adapted from Flint, 1930, figure 36).

A map of Wethersfield provides a good example of the colonial layout of an early-seventeenth-century riverine village. The fields lay side by side and were densely clustered in the town center (fig. 4–8). This might have been the reason early settlers continued to experience significant problems raising enough food; planting monoculture crops in wide-open fields situated close together encouraged pests, whose populations skyrocketed following this great windfall. The most serious problem farmers faced, however, was that European grain was difficult to grow in New England's acidic soils. Regularly spreading lime and fertilizing with manure were essential for raising healthy crops, but both were initially in short supply. One disillusioned farmer wrote that after five or six years, the soil became exhausted and grew barren beyond belief. Despite these problems, however, Connecticut colonists were generally well off. Historian Howard Russell wrote that by the middle of the seventeenth century the "average" farmer owned four cows, three goats, and eight hogs while working five acres of Indian corn, an acre of wheat, and an acre of oats/barley and peas. Thomas Nowell of Wyndsor left an estate worth 296 pounds and 17 shillings when he died in 1648, indicating that some accrued significant wealth in a short time (Table 4-1).

A serious threat to English crops arrived from the Old World in the 1660s: a fungus that caused black stem rust, or the "blast." The disease destroyed the leaves of wheat and rye before seed production. In addition, overgrazing and plowing caused rapid soil exhaustion and erosion. Both became endemic to colonial agriculture. Lands cleared for crops frequently had to be turned back to pasture or woods in less than a decade.[6] The introduction of English grasses had succeeded by the early 1650s, helping Springfield become the first major livestock-producing area in colonial New England.

Over time, the Puritan settlers became less dependent on their native neighbors. When the fur trade in the northern reaches of the Connecticut Valley collapsed after 1655, the natives began to obtain European goods on credit, using their land as collateral, and as a result more and more land became available to the settlers. Expansion of the colonial settlements in the riverine section of the Central Valley slowed temporarily because of conflicts with the natives that climaxed with King Philip's War in 1675, and Deerfield and Northfield remained frontier towns until well into the eighteenth century. Despite those problems, the production of both wheat and livestock increased rapidly, and the Central Valley became the "bread basket of New England," supplying New York City and Boston.

6. Almost a century passed before valley farmers began to rotate crops, which reduced soil exhaustion and decreased the occurrence of the blast. By the late eighteenth century, a typical cycle would include wheat followed by beans, then corn—or leaving the land fallow—followed by rye or barley, and finally clover.

Table 4–1. Inventory of the Estate of Thomas Nowell of Wyndsor, Deceased February 22, 1648.

	Value in pounds/sh
The dwelling house, barne, outhouses and homelott orchyard;	75
13 akers of meadow	45
66 akers upland	03
2 horses, one colte	27
2 oxen, 2 steares	23
3 cowes, 1 heifer, 1 bull	18.05
3 swyne	02
Abroad, 2 cowes, one steare	15
4 stocks of bees	03
In the seller; 2 pounds porke	03.10
7 bush rye, 3 bush malte, 2 bush pease	04.13
22 bush wheat	04.08
14 yards okam cloth	00.18
In the parlour; 1 standing bed with its furniture	17
1 trundle bed with its furniture	10
A cubbered, a table, a chaire, a small box, 3 stoole	02.10
2 trunks, 1 chest, 14 cushions	03.12
3 table cloaths, 15 table napkins	02.18
14 yds ½ new linen, with some cotton cloath	02.03
And. . . .In money and plate	34

Walter Hard, 1947, pp. 91–92.

During the American Revolution, George Washington referred to Connecticut as the "Provisions State." Whether he was referring to the more than thirty thousand Nutmeggers capable of bearing arms or the state's abundant food supplies is unclear. Most likely, it was both. Between 1768 and 1772, Connecticut led the colonies in exports of barreled meat and livestock. During the bitter winter at Valley Forge, Washington made a desperate plea to Connecticut for cattle in order to prevent the dissolution of the army. Governor Jonathan Trumbull ordered immediate purchases, and large herds were driven to Valley Forge. The starving revolutionaries devoured the first group of animals in just five days.

In the decades following the Revolution, Connecticut rapidly progressed from an internally driven economy, consuming most of its agricultural produce and importing relatively little, to an active trading community. By 1790, Connecticut supplied two-thirds of all the live cattle shipped abroad. Although those exports were important, they did not significantly influence the general American economy until a century later, because the trade primarily involved an exchange of goods: horses and wheat for salt, sugar, and rum.

John Adams considered the late eighteenth century a time of great progress. While traveling on horseback from Windsor to Middletown in June 1771, he remarked, "Nothing can exceed the Beauty, and Fertility of the Country. The Lands upon the River, the flatt low Lands, are loaded with rich, noble Crops of Grass, and Grain, and Corn." A citizen of Wethersfield told Adams that "some of their Lands, yielded 2 Crops of English Grass, and two Ton and a half [of wheat and corn] at each Crop, and plenty of after feed besides—but these must be nicely managed and largely dunged."

According to some estimates, more than 80,000 people lived in the low-lying areas along the coast and in the Central Valley of Connecticut and Massachusetts in the late eighteenth century.[7] Thomas Poronall wrote that the land between New Haven and Hartford was "a rich, well cultivated Vale, thickly settled and swarming with people. . . . It is thought you were still traveling along one continued town for 70 or 80 miles on end." The alluvial terraces west of the river from Middletown to Windsor had become this continent's first wheat belt.

In his characteristic tongue-in-cheek style, Odell Shepard wrote, "In early days our people could see the lights in one another's windows and could communicate by shouting from farm to farm. An old man has told me that in his youth it was possible to arrange a barn dance among a dozen of his neighbors without any man's stirring from his front door. Another old man has told me of the time, far back, when he, sitting under his elm tree on a Sunday afternoon, could see seven or eight friends sitting in front of their several houses and under their own elms." By 1820, Connecticut had too many farmers for the extent of its territory, and people began to look to immigration and/or industrialization.

7. Population density clearly shows the economic significance of the geologic framework and soils of the Connecticut Valley. In 1756, an average of forty-five people occupied each square mile of land in the valley; during the same period, only eight inhabited the rugged highlands.

The Industrial Revolution

The highly successful nineteenth-century industrial revolution in Connecticut was mainly due to four factors: the availability of startup funds, the ore deposits, hydropower, and last but certainly not least, Yankee ingenuity and manpower.

Citizens of towns and cities in the Central Valley, who had made their wealth exporting agricultural goods, broadened their investments to include small industrial enterprises. The machine-tool industry originated in Old England but was developed and refined in New England. Textile mills rose along rivers in the eastern highlands, and blast furnaces crowded streams in the northwest. Towns rapidly grew around those industries and new economic centers developed, both of which helped spread wealth more uniformly across the state.

Connecticut artisans and inventors took out patents at nearly three times the national rate. Entrepreneurs such as Chauncey Jerome, Eli Terry, and Seth Thomas (clocks), Samuel Colt and Simeon North (firearms), Elisha Root and Eli Whitney (machinery), Samuel Morey and John Fitch (steam-propelled boats), and Charles Goodyear (rubber tires) devised machines and products that were adopted throughout the United States and even exported to Europe. Those industries laid the foundation for manufacturing eminence, and soon Connecticut became known as the "Industrial Incubator."

In 1902, while residing in Washington, D.C., William Countryman, a former editor of the *Hartford Post*, detailed the state's manufacturing preeminence:

> At my boarding-house I find the plated ware to be of Connecticut manufacture. The clock that tells me the time from the mantelpiece; the watch my friend carries; the hat he wears; his pocketknife, are all from Connecticut. At the office I write with a Connecticut pen and when I need an official envelope I find that the original package from which I take it bears a Connecticut mark. If I make an error and wish to erase it, I do so with a steel eraser made in Connecticut, and my letter finished I deposit it in a corner letter box, stamped 'New Britain, Conn.' This letter, I am sure, when it reaches its destination, is delivered from a post-office box locked with a Yale key. My desk has a Connecticut lock and key although perhaps made in Michigan. In looking about

> the city I am attracted to a shop window glittering with swords, and read on an ugly looking machete this inscription: 'Hartford, Conn., U.S.A.' A Winchester or a Marlin rifle, or a Colt's revolver, all made in Connecticut, I find in another window, and in still another supply of fixed ammunition from New Haven and Bridgeport. Axes, hammers, augurs, all kinds of builders' hardware, and in a shop close by—all made in Connecticut. Foulards, cottons, woolens, worsteds, rubber goods of all kinds, are near by—they are standard makes from Connecticut. The gas and electric fixtures that turn them off are of our manufacture, I doubt not. Do I want a button? Made in Connecticut. 'Hand me a pin.' The box tells me it is from 'Waterbury, Conn., U.S.A.' That automobile rushing by came from Connecticut. That bicycle, those tires, these novel call and door bells—all from Connecticut. Typewriters on every side from our little state. And if I lounge through residential streets summer evenings, I hear from many open doors and windows the sound of music. This may not be from a Connecticut piano, although in most cases the ivory keys would be found to have been made in our state, but in many instances emanates from a Connecticut-made gramophone or phonograph. And what of the sewing machine? Everybody knows that the earliest ones were made in Connecticut, and the latest improved are made there now in great numbers. And last let me say that where my trousers are put away at night they go into a hanger of the best kind—made in Connecticut.

In 1939, Odell Shepard asked a rhetorical question: "How was it that those old Connecticut Yankees invented half the things that run the world?" The reply appears to be that such natural resources as hydropower, Central Valley soils, Salisbury iron ores, Bristol copper deposits, Portland brownstones, and Stony Creek and Westerly granites triggered Connecticut's industrial revolution. It evolved, however, because of Yankee ingenuity and the invention of interchangeable manufacture.

The Metacomet Ridge

The Scientific, Political, and Cultural Impact of an Old Lava Flow

Before the time when Connecticut acquired
a reputation as "the land of steady habits,"
—in the days, perhaps, of its wild oats, so to
speak, it was the field of volcanic activity of
far greater power than has ever been
manifested by Vesuvius.
—William Pynchon, 1896

Anyone familiar with the devastation in Pompeii and Herculaneum caused by the 79 CE eruption of that mighty volcano will find Pynchon's statement to be a stretch. After all, there is no sign of a volcano in Connecticut, not even the deeply eroded remains of one. However, Pynchon was right: volcanism destroyed much more territory here than the infamous Vesuvius ever did in Italy.

Connecticut's volcanic remains occur in a long, narrow ridge that rises as much as five hundred feet above the adjacent lowlands and dominates the topography of the Central Valley. The ridge was named after Metacomet, also known as Pometacom or King Phillip (fig. 5–1), a sachem of the Wampanoag peoples who played an important role in an especially violent native uprising against the settlers. Legend has it that he and his followers spent time in a cave in the northwestern corner of Talcott Mountain, a segment of the ridge, from which they could control part of the Farmington valley.

Fig. 5–1. Metacom (Pometacom) or King Philip, after whom the basaltic backbone of the Central Valley was named. The different "portraits" show the stereotypical views about Connecticut natives commonly held by its residents in the early (left) and late (right) halves of the 19th century.

The ridge is exposed inside a rift valley and stretches almost unbroken from Amherst to Meriden. In the Meriden area, however, a swarm of north-northeast-trending faults offsets the hills to the east. Because of significant vertical motion along these faults, the ridge broke into segments, or fault blocks, each with its own name: South Mountain, Cat Hole Mountain, Lamentation Mountain, and Chauncey Peak. From Mount Beseck in Middlefield, the Metacomet Ridge can be traced farther south to East Haven (plate 2, bottom and fig. 5–2).

The Metacomet Ridge is the erosional remnant of a 500- to 600-foot-thick lava flow, a slab of rock that filled the entire Central Valley around 200 million years ago.[1] The flow, named after the town of Holyoke, tilted to the east tens of millions of years later, together with the entire contents of the rift zone. As a result, the ridge became asymmetric in profile and now exhibits precipitous western and gently dipping

1. Assuming the floors of a building to be ten feet high, the flow's maximum thickness is equal to that of a sixty-story skyscraper, a little more than half the height of the Empire State Building.

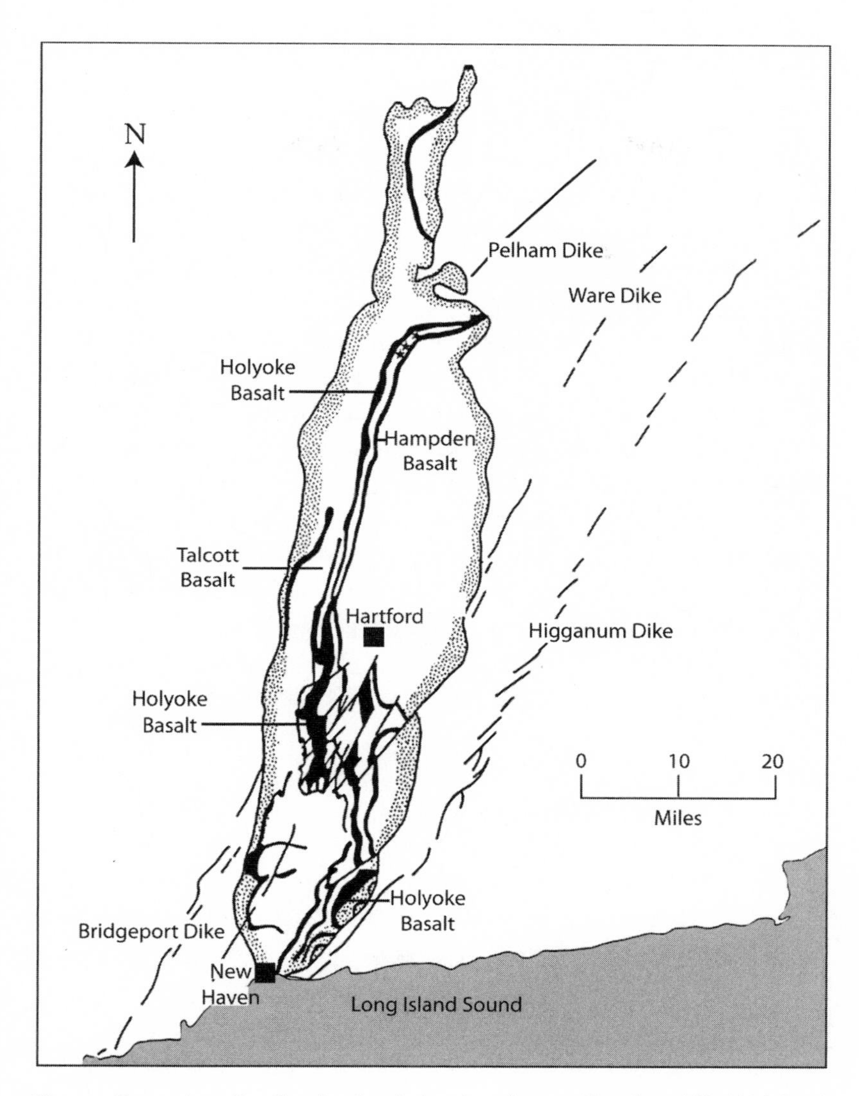

Fig. 5–2. Connecticut rift valley showing the location of exposed lava flows/sills (thick lines). The Holyoke flow was predated by the Talcott and followed by the Hampden basalt. Note location of New Haven and Hartford on different sides of the Metacomet ridge. Thin lines in the Highlands represent possible feeder dikes.

eastern slopes. Segments of the flow west of the ridge eroded and became part of the sediments deposited offshore, while younger sandstone formations bury the flow east of the ridge.

Massive sheets of ice that scoured Connecticut in the more recent geologic past greatly accelerated the erosion of the Metacomet Ridge. Steep northeast- and northwest-trending clefts developed where ancient faults cut the lava sheet. They broadened into gorges when the ice pushed through and removed the broken-up basalt. The lower slopes of the steep western flanks of the ridge are typically coated with talus, large aprons of loose basalt blocks that grow each winter when the process of ice wedging detaches rocks from the cliff face.

The ridge remained undeveloped for centuries and presently forms a scenic green belt in the densely populated Central Valley, a seemingly untouched wilderness that separates urban and suburban agglomerations on either side. Although obviously quite different topographically, this strip of land could be referred to as Connecticut's Central Park.

John Whittier, a well-known poet, greatly admired the ridge and wrote the following verse in 1838, after visiting Daniel Wadsworth's property, "Monte Video," on the crest of Talcott Mountain, west of Hartford.

> Beautiful mount! With thy waving wood,
> And thy old gray rocks like ruins rude
> I love to gaze thy towered brow
> On the gloom and grandeur and beauty below
> When the wind is rocking thy dwarfish pines,
> And the ruffled lake in the sunlight shines

Not surprisingly, its scenery became a focus for paintings and legends as well.

Because of its topography and special type of rock, the ridge's ecology is unique and rather diverse. Its summits are warm and dry; the black rock greedily absorbs solar heat, and the thin soils retain little moisture. This creates habitats with affinities to drier ecosystems further south. Cool conditions prevail on the talus slopes at the base of the ridge, where the ice that had accumulated between the large rock fragments in the wintertime creates an icebox effect in the summers. The soils are thicker and moister, providing boreal habitats similar to those of more northern ecosystems (plate 9).

Geologic Setting

About 230 million years ago, the supercontinent of Pangaea (fig. 5–3), an accumulation of most of the continents on Earth, began to break apart along former suture zones in the Appalachian Mountain belt. As the North American and African continental masses detached and began to drift apart, the crust between them stretched, reactivating many old fractures. Vertical slippage along such faults caused intermittent subsidence of elongated segments of crust and the formation of long, narrow basins, or rift zones. The climate in Connecticut was subtropical but harsh; long dry periods alternated with relatively short wet periods, during which rapidly flowing streams swept erosional debris from adjacent highlands into the basins.

New sets of north-northeast-trending fractures broke through the brittle crust about 200 million years ago (see fig. 5–2) and tapped large reservoirs of molten rock at depths of about seven miles. Those magmas

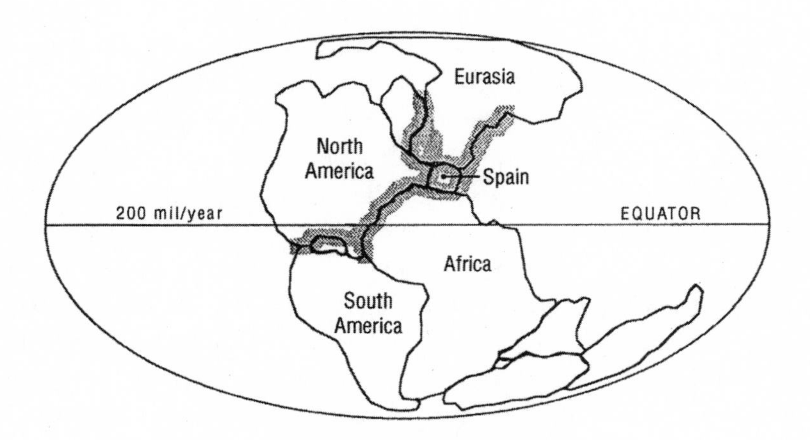

Continental Configuration About 200 Million Years Ago

Fig. 5–3. Plate tectonic configuration around 200 million years ago, when Pangaea began to break apart and Connecticut's Central Valley formed. Voluminous batches of magma erupted in the fractured (shaded) zones. Note approximate location of the equator (adapted from Olsen et al., 1992, figure 1).

used the newly created pathways to surge to the surface, where they spread out across the sediments, filling the valley from margin to margin. The magma that reached the surface emerged from long, gaping fissures and developed into thick sheets with the consistency of molasses that inched across the sun-baked red soils. Clouds of water vapor and poisonous gases whistled from cracks in the lava crusts, blanketing the land and polluting the atmosphere. We might envision it as follows:

Early one morning, the light that precedes sunrise crept across the desert sand and brightened its russet hue. When it reached the bottom of a steep cliff, the light woke the small herd of Anchisaurus dinosaurs that had drawn together for protection against the chilly night winds. The lead female rose, sniffed the air, and headed for a hazy white ribbon on the horizon, the signature of a shallow lake. The herd followed her along the dry streambed of an arroyo. They were barely halfway to the enticing source of water when the ground shuddered. Confused and agitated, they hastily gathered the young that had trailed behind and waited. When the land came to rest, the animals continued their trek east.

When they reached the swampy shore and quenched their thirst, they left their three-toed tracks in the shallow mud (fig. 5–4). Suddenly, a second, more powerful earthquake struck. The ground buckled and heaved, and a segment of the cliff they had left a few hours earlier collapsed in a cloud of dust. Gaping chasms opened at the foot of the scarp above a buried fault that had ruptured. Before long, fountains of hot ash soared into the air from a linear series of vents, rapidly forming a row of volcanic mounds. Steam billowed into the morning sky, and waves of poisonous gases, heavier than air, flowed downhill toward the lakes in the center of the basin. The herd scattered in panic, making for deeper water, but the gases overtook the fleeing animals, and one after another they collapsed.

By nightfall, sheets of magma began to issue from the fractured earth, and a noisy steaming and crackling wall of hot lava slowly oozed across the land. When it reached the lake, the lava vaporized the water and everything in it. Of the dinosaurs, only their tracks remained—baked solid. In the following days and weeks, more and more magma surfaced, finally filling the entire valley and turning it into a huge graveyard of steaming stones, a true inferno.

Afterward, the volcanic scene remained desolate for thousands of years, but the floor of the rift valley continued to subside slowly, and fresh

Fig. 5–4. Dinosaur track, Wesleyan University.

sediments washed in, filling the depressions between the lava blocks and gradually covering the flow itself. Tiny green islands emerged inside the black expanse. Over time, ponds and small lakes formed, and the descendents of surviving dinosaurs returned to the valley.

Three times in a period we assume to be less than 600,000 years, voluminous batches of magma reached the surface, and lava flows with temperatures of more than two thousand degrees filled the valley. The earliest volcanic event left the Talcott flows, composed of several sheets up to 190 feet thick that thin out toward the north. The second event created the Holyoke flows, two sheets with a combined thickness of up to six hundred feet. The volume of the Holyoke flows is estimated to be close to two hundred cubic miles but might have been much greater, if the lavas extended onto adjacent highlands, as some geologists believe. The third volcanic event left the Hampden flow, which is "only" about 180 feet thick.

Large volumes of magma never reached the surface but cooled at depth inside the faults, forming subvertical dikes, or found themselves squeezed between the horizontal layers of older sediment layers, giving birth to large, blister-like bodies (sills), such as the Sleeping Giant north of New Haven and the Barndoor Hills west of Granby.

Intrusions of magma along feeder dikes and their emplacement as flows in the Connecticut Valley were by themselves truly catastrophic events for southern New England. Volcanism, however, was not restricted to Connecticut and Massachusetts. Similar basins formed throughout the Appalachians and in western Africa, Spain, and northern South America. The Fundy basin of Nova Scotia and the Newark basin in New Jersey are exposed rifts that are closest to Connecticut. The latter contains the famous Palisades sill, a pillared intrusive mass exposed on the western shore of the Hudson River opposite New York City.

Basaltic magmas contain volatile elements that combine as water vapor, carbon dioxide, and sulfurous gases. As the magma cools, crystals form, and the gases are driven out into the atmosphere, where they affect regional weather patterns. When emitted in very large volumes, they have changed the global climate for years. Benjamin Franklin, who represented the United States at the court of Louis XVI, complained that the sulfurous haze hanging over Paris in 1783 gave him coughing fits and stung his eyes. He correctly attributed the foul air and anomalous cold of the following winter to an eruption of a volcano on Iceland.

In 1783 and 1784, magmas that had erupted along the Laki fissure in Iceland flowed out over the adjacent plains. The total volume of the flows was estimated at only three or four cubic miles, yet the accompanying gases affected the weather in much of the northern hemisphere. Clouds of carbon dioxide hung in the lower atmosphere and spread across the Atlantic toward England, and their greenhouse effect caused unusually warm summer months. Sulfur dioxide rose above the clouds and combined with water vapor, creating veils of tiny sulfuric acid droplets that spread across northern Europe. The veils in turn reflected some of the incoming solar energy and caused an unseasonably cold winter. Needless to say, the people of Iceland suffered the most. Dense clouds of noxious, heavier-than-air gases created a bluish haze that filled the valleys, poisoning the vegetation and killing—either directly or indirectly—half the cattle and three-quarters of the sheep. The "blue haze" famine ensued, and more than a quarter of the Icelandic population succumbed.

The volume of the 1783–84 flows was about fifty times smaller than that emplaced in the Central Valley during the Holyoke event alone. No doubt exists, therefore, that the latter eruption would have affected the weather in the entire northern hemisphere for several years.

The cumulative volume of all of the basaltic magmas that surfaced in the Appalachian rift zones is several thousand times greater that of the Laki event. The three flows in the Connecticut and Newark rift valleys appear to have been more or less contemporaneous, but they could have preceded or followed the Nova Scotia outpourings by many thousands of years. Most likely, the climate deterioration caused by such intensive volcanism occurred intermittently over a period of almost half a million years. Its cumulative effect on the global climate resulted in major extinctions of both land and marine organisms and probably caused a bottleneck in the evolutionary trend of the early dinosaurs. The ecological consequences were so significant that they led paleontologists to recognize the end of one geologic period, the Triassic, and the beginning of another, the Jurassic.

Pluto versus Neptune

In the late eighteenth century, when the earliest studies were made of rocks presently known as volcanic, a heated debate raged in Europe, between the "Neptunists," named after the Roman god of the sea, and the "Plutonists," named after Pluto, the Roman god of the fiery underworld. The Neptunists were headed by the Prussian geologist Abraham Werner, the Plutonists by a British geologist, James Hutton, and a French bureaucrat, Nicolas Desmarest.

According to Werner and his followers, all of the rocks on Earth had originated in universal oceans as deep as our present-day mountains are high. They believed that global flooding had occurred twice—first during the Primeval Chaos and again during Noah's deluge. When the biblical floodwaters receded, the ocean floor deposits were gradually exposed. Werner recognized two types of rocks—those that had formed from debris washed in from emerging dry lands (the sediments) and those that had been chemically precipitated below sea level (the basalts), the way salt forms in basins with evaporating seawater.

Desmarest, a civil servant who had traveled widely in France and studied the extinct "volcanic" landforms of the Massif Central, concluded that the formations he had encountered there were the products of molten rock that had risen from the Earth's interior and cooled above sea level. Hutton had studied the remains of an extinct volcano near Edinburgh as well as the large domes of granite found in Scotland. He was convinced that an internal source of heat would be required for their molten origin, which meant that they had to be Plutonic.

While debates between the two camps continued in Europe, two Yankees tried to resolve this problem in America. Professors Benjamin Silliman of Yale and Edward Hitchcock of Amherst College had both studied the basaltic layers of the Central Valley and had become convinced of their Plutonian origin. The men's Puritan backgrounds, however, made it difficult to go against church teachings. In 1810, after a visit to England in 1805–6, Silliman wrote:

> Arthur's Seat, at Edinburgh [the remains of a volcano], exhibits regular six-sided prisms, and our rocks here [in Connecticut] show a similar tendency so strongly, that one would, from this circumstance alone, be induced to suspect their identity . . . There can be no hesitation in pronouncing [the rock] to be what is called whin stone in Scotland, trap in Sweden, and basalt in some countries . . . after comparison with the trap rocks in the vicinity of Edinburgh, I felt assured of their igneous origin . . . I felt greatly relieved when I was excused from attempting to compel myself to believe that porphyry, trap in all its varieties, and even granite, had ever been dissolved in water. I became, therefore, to a certain extent, a Huttonian.

However, in his lectures at Yale Silliman continued to hesitate, linking geological observations to biblical accounts and insisting that every science had its own metaphysics. Silliman told his students that "knowledge is nothing but the just and full comprehension of the real nature of things, physical, intellectual, and moral; it is coextensive with the universe of being, both material and spiritual; it reaches back to the dawn of time, and forward to its consumption; nay, it is coeval with eternity and is inseparable from the incomprehensible existence of Jehovah. Only one mind therefore intuitively embraces the whole."[2]

2. Silliman might have read William Paley's *Natural Theology* during his sojourn in England. In this book, published in 1802, Paley argues that religion and science should not be considered separate, that their concepts could be combined using reason.

Early on, Hitchcock, a geologist and clergyman, took an unambiguous stance on the origin of the basaltic flows in Massachusetts but hesitated to "come out." In an anonymous paper he wrote in 1824, Hitchcock concluded: "The resemblance of the amygdaloidal traps to certain varieties of lava; the convulsion and distortion of other strata [sediments] in their vicinity; the change in other rocks in connection to trap dykes [answers] precisely to the action of heat . . . To what, but a volcanic agency, can such circumstances lead the mind?" Hitchcock did not claim authorship for his forward-thinking paper until 1863, more than forty years after its publication, indicating the force of clerical opinion among New England academicians.

Hitchcock's and Silliman's hesitance to promote their theories is the more surprising because as early as 1821, Thomas Cooper, president of South Carolina College, published a paper in the *American Journal of Science* in which he argued convincingly that "the basalt formation of Werner and his followers was of igneous origin and came under the head of volcanic ejections."

To clinch the argument, Silliman made a detailed study of the contact between the Hampden flow and the underlying sandstones at Rocky Hill (fig. 5–5). In 1830, he finally bit the bullet and concluded that "the sandstone was evidently formed by subsidence, from mechanical suspension in water and is composed of the accumulated ruins of other rocks . . . the trap was as evidently deposited and aggregated, not from mechanical suspension, but from a state of chemical mobility . . . we must I think without doubt admit its [the basalt's] igneous origin, we must not hesitate to go where truth and evidence, and sound reasoning will carry us."

This publication represented a major step forward in the recognition that volcanic formations existed in Connecticut and elsewhere in the Appalachians. However, in Europe it took more time before theories relying on scientific observations replaced Werner's biblical hypotheses.

Some Yankees remained on the sidelines, even trying to be both Neptunists and Plutonists. In 1836, John Barber, a well-known artist and Connecticut historian, wrote: "It is now believed by all geologists that the materials composing the trap rocks were melted in the bowels of the earth and thrown upwards through the incumbent strata by igneous action, and that the peculiar formation of these rocks . . . was

Fig. 5–5. Contact of sandstone (brownstone) and overlying lava flow in Rocky Hill quarry (Silliman, 1830; frontispiece attributed to Daniel Wadsworth).

caused by the pressure of water from above, it being unquestionable among geologists, that our globe was once covered with a deep ocean."

Drifting from the Tropics Northward and Westward

From a mineralogical viewpoint, Holyoke basalt is a relatively simple type of rock. It is composed mainly of silicates, light-colored plagioclase feldspar crystals, rich in calcium, and dark-colored augite crystals that are rich in iron and magnesium. Together, these minerals give the fresh rock a salt-and-pepper appearance. Black magnetite crystals occur here and there within the chaotic framework of these minerals and retain a memory of Earth's magnetic field, which was present when the magma cooled. In the mid-1960s, I used this information about that ancient magnetization to determine the geographic location of Connecticut about 200 million years ago.

Earth's magnetic force lines vary from vertical at the poles to horizontal near the equator (fig. 5–6). Therefore, a relationship exists between the inclination, or tilt, of these lines and their geographic latitude. The current magnetic field in Connecticut is represented by a magnetic force that points north (fifteen degrees west of true north), and dips steeply (about seventy degrees) into the ground. The fossil magnetization measured in the Holyoke lava flow also points north but dips at a relatively low angle of about twenty degrees. The lower dip means that the emergence and cooling of the magma must have occurred when future Connecticut was located much farther south, closer to the equator, which at that time crossed present-day Florida (fig. 5–3). When the Atlantic basin opened, the North American plate must therefore have not only moved west, separating from Africa, but also moved north, changing Connecticut's latitude by about twenty degrees.

The warmer climate at the lower latitude supported different life forms, and fossil evidence for many animals and plants exists in the old sediment layers below and above the lava flows. Most impressive and unique are the fossil dinosaur footprints exposed at Rocky Hill. These tracks survived because the sun baked the original prints, made in wet clays, to the consistency of red brick, and the runoff that followed later

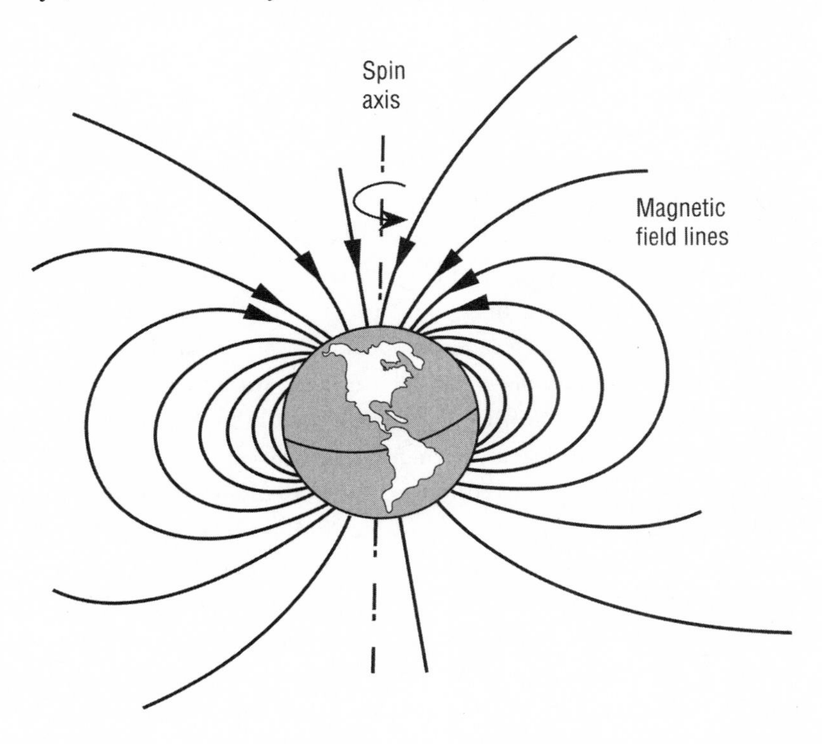

Fig. 5–6. The Earth's magnetic field lines, which are horizontal at the Earth's surface near the equator, vertical near the poles. When this magnetization is incorporated into rocks, it provides evidence for the latitude at which they formed.

downpours covered the tracks with new layers of sand and clay, protecting them from erosion. At Rocky Hill, a number of prints are arranged along clear track ways made along an ancient lakeshore. Imprints of raindrops inside some of the tracks indicate that at times the animals walked in the rain (singing or not).

Economic Significance of Basalt Flows

When magma cools, its volume shrinks, and numerous fractures develop in lava flows. As cooling proceeds, the fractures grow and often form polygonal columns (fig. 5–7). The magma in the central part of the flow remains hot for longer periods than elsewhere, and there the cooling joints form more chaotic patterns. The columns are commonly densely packed and range in diameter from a few inches to about a foot. Although the separations between adjacent columns are narrow, they are numerous, providing the flow with plenty of space to hold water. The fractures also connect, making the lava flow very permeable and allowing water to move through the network of fissures. Acting like a giant sponge, it soaks up and retains rainwater and then slowly releases it to adjacent valleys, ameliorating the effects of pronounced dry seasons.

Because the sandstone formations below and above the flows eroded more easily, many wetlands and ponds formed naturally adjacent to the ridge.Their bottoms are lined with glacial clays that increase their water-holding capacity. Because of the mounting demand for water in the Central Valley, a number of these wetlands were turned into reservoirs for cities such as New Haven, Middletown, Hartford, and Springfield. Currently, there are more then thirty lakes and reservoirs on either side of the Metacomet Ridge in Connecticut alone. The largest is Lake Gaillard, which supplies clean water for New Haven.

The sponge effect of the basalt is especially evident where fragments have broken away from the cliff faces and accumulated in debris fans. Large volumes of water can concentrate in these zones, leading to the formation of natural "icehouses." The most spectacular was located in the Cat Hole Gorge, north of Meriden. This once precipitous defile trends north to south and was so narrow that the sun's rays could penetrate to its bottom for only an hour or so, and then only when the sun was near its meridian. Rainwater that accumulated in the debris fans

Fig. 5–7. Outcrop in Holyoke basalt. Note the numerous fissures and columns formed during cooling (and shrinking) of the lava flow.

froze during the winter and provided a stream of ice-cold water throughout the summertime.

Water companies have bought large segments of the Metacomet Ridge to protect their reservoirs. In the last decade, however, developers have become increasingly interested in constructing homes (with a view) on the remaining parcels. The septic wastes and runoff from those houses and their driveways/streets pose a threat to this unique basalt aquifer and its associated reservoirs.

Basalt, commonly known as traprock, is the most valuable commodity in the valley and continues to be quarried on a large scale. In several areas, this activity has taken big bites out of the Metacomet Ridge. Over time, more than twenty basalt quarries have developed, and several remain active. The rock is used in roadbeds, concrete aggregates, and septic systems. Its toughness is believed to have been a major stimulus in the development of the prototype rock crusher that Eli Whitney Blake of New Haven invented in 1858. Similar crushers continue to be in use to this day. The largest basalt quarry is located inside the Totoket Mountain in North Branford; its exposed rock face extends for more than a mile.

The Political Divide among Canalites and Riverites

The principal colonial population centers in the Central Valley of Connecticut were New Haven on the western side and Hartford on the eastern side of the natural wall formed by the Metacomet Ridge (see fig. 5–2). In addition to their topographic separation, these towns were settled by colonists with different backgrounds. According to Cotton Mather, the founders of the New Haven colony were of urban origin: "Londoners and merchants. Men of traffic and business, their design was in a manner wholly to apply themselves to trade." The founders of the three settlements on the Connecticut River that comprised the Hartford area had rural backgrounds. Most were trained in or had experience with agriculture, and their principal aim was to make a living by farming. Those differences, and growing religious discord, led to a philosophical schism that their physical separation then nurtured further.

New Haven joined the Connecticut Colony in 1665 and soon competed with Hartford both commercially and politically. For the next two centuries, the two towns alternately served as the capital, which required legislators to travel back and forth regularly. New Haven, a major East Coast deepwater port, was ideally located for the import of goods but lacked a hinterland from where significant volumes of exportable goods could be produced. Hartford possessed an extensive hinterland that stretched deep into Massachusetts and produced a wide array of agricultural products. However, rapids, tidal changes, and shifting shoals at the mouth of the Connecticut River made transporting those goods by river difficult.

The Metacomet Ridge made travel over land between the towns uncomfortable and the shipment of goods very cumbersome. There were only a few turnpikes, and they were all muddy or rocky and deeply rutted. Bridges and culverts were scarce, which made crossing streams hazardous. Stagecoaches connected the towns three times a week, but almost all travel ceased during the winter months, when snowdrifts buried the roads.[3]

For decades, the leading citizens of New Haven had looked with envy on the prosperity that river traffic brought to Hartford and Middletown. Because no major river flowed into their harbor, New Haven's citizens believed that their businesses would remain local in character and limited in extent unless they could unite with towns farther north by artificial means. The construction of the Erie Canal and its early economic successes showed how a waterway could significantly increase a town's economic base.

By the turn of the century, the economic situation for New Haven had become worse because of the bypass canals constructed around rapids and falls in the Connecticut River, including Millers and Montague Falls (Massachusetts) in 1800, Bellows Falls (Vermont) in 1802, and Sumner and Olcott Falls (Vermont) in 1810. These locks extended the reach of commercial enterprises by more than a hundred miles, greatly benefiting the major river towns.

James Hillhouse, a respected state senator and citizen of New Haven, recognized the benefits of a canal connecting New Haven with Northampton, Massachusetts, bypassing Springfield, Hartford, and Middletown. Hillhouse found support for his plan, and Benjamin Wright, the chief engineer for the Erie Canal project, was hired to make a survey of the proposed canal route. In a letter to Isaac Mills, Wright wrote that this was to be "the first project of its kind in Connecticut carried into effect and would be the incipient step to works of internal improvement that [would] be a lasting monument to the

3. Stagecoach service between Hartford and New Haven began in 1771, and the state's first turnpike was constructed in 1792.

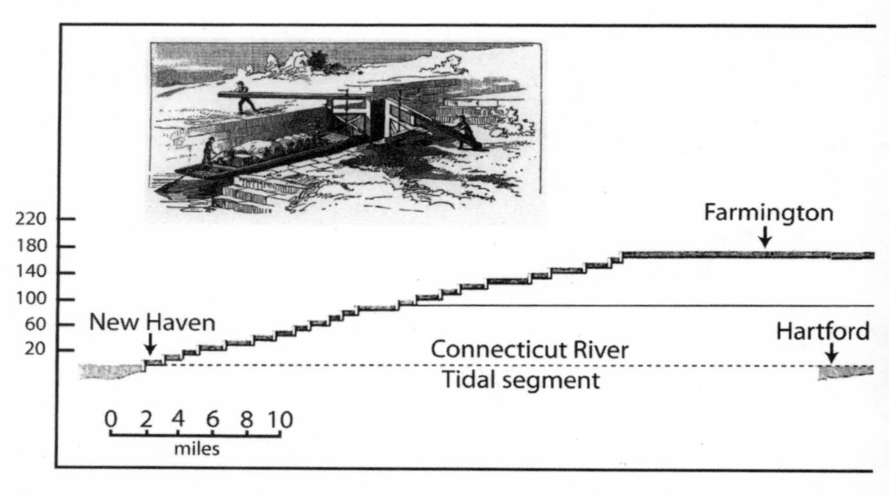

Fig. 5–8. Architectural drawing of the Farmington and Enfield canals in cross-section (Hurd, 1828). The Enfield canal was finished in 1829, the Farmington canal in 1835.

enterprise and intelligence of high-minded people." Wright estimated the cost (quite precisely) at $420,698.88, though the canal would ultimately cost investors more than two million dollars. Even in those days, overruns were quite common.

The citizens of Hartford decided to counter New Haven's threat and proposed "to lock the falls at Enfield" and "construct a dam or dams for the purpose of entering and leaving the locks in still water." With the incorporation and organization of the Connecticut River Company, a prolonged and bitter controversy began between Hartford's "Riverites" and New Haven's "Canalites." The Canalites attempted to block the Riverites' efforts to procure the requisite authority for building a canal that would bypass the Enfield rapids. The Riverites tried to defeat the petition for a charter to extend the New Haven canal across the border to Northampton. The crucial battle took place in the General Court of Massachusetts. After two years of discussions, the court removed the subject from its domain by granting both petitions. The race to complete the rival projects was on.

The canal bypassing the Enfield rapids was completed in 1829. Chartered in 1826, the New Haven–Northampton canal took thirteen years to survey, design, and construct. It ran north through Farmington and Southwick to Northampton, where it connected to

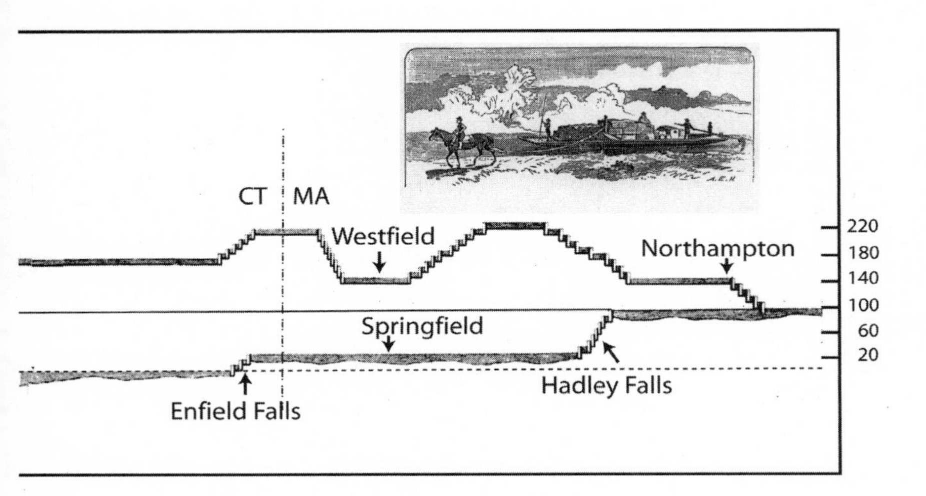

the Connecticut River. Using picks, shovels, and wheelbarrows, workers dug a canal thirty-four to thirty-six feet wide, about six feet deep, and eighty-seven miles long into the unconsolidated glacial debris and weathered sandstone of the Central Valley west of and parallel to the Metacomet Ridge. The workers reached Cheshire in 1826, Farmington in 1828, and the Massachusetts border in 1829. Finally, on July 4, 1835, the first boat docked in Northampton. Church bells rang, cannons pealed, bands played, and festive crowds rushed to the canal.

Sixty locks were needed to lift a northbound boat 310 feet and then lower it 213 feet to reach the Connecticut River near Northampton (fig. 5–8). In addition, bridges and aqueducts had to be constructed. The largest aqueduct consisted of a 280-foot-long wooden trough supported by six stone pillars that crossed the Farmington at thirty-six feet above river level. A similar structure crossed the Westfield River.[4] The canal boats were approximately the size of present-day tractor-trailers but drew only four feet of water. Teams of two or three horses pulled the boats, covering the eighty-seven miles in about twenty-five hours. In 1838, steamboats were allowed on the canal. The

4. Detailed maps of the Farmington (Connecticut segment) and the Hampshire & Hampden (Massachusetts segment) canals were produced by Carl Walter in 2000 and 2006, respectively.

trip from Northampton to New Haven with these boats took only twelve hours, at a cost of $3.75.

From the start the canal had major problems. The editor of the *Northampton Courier* wrote on June 7, 1837, less than two years after the canal opened: "The Canal disappoints every expectation. Half the time, when the locks don't squash in and the banks but half burst out and the bridges only shiver in the wind, the boats get stuck in the mud and the owners are obliged either to work their passages on shore or hire people to dig them out."

The canal leaked because of its shoddy construction and the use of porous clays for its bed. To supply sufficient water to the canal and its locks, several streams had to be diverted. The Farmington River was dammed at Unionville and its waters fed into the canal. Low water in dry summer months and problems with ice in the winter reduced the canal's effectiveness. A drought in 1845 closed the canal from July through September. In addition, there were occasional incidents of sabotage by farmers whose land and water rights had been seized.

In 1845, railroad tracks were built alongside the canal; within a year, most people preferred to travel by rail than by boat. In 1848, more than twenty years after the work started, the New Haven and Northampton Company went bankrupt. A costly economic venture with strong political overtones had failed. Hartford had won and became Connecticut's full-time capital in 1875.

The Metacomet Ridge thus played a major role in solving an academic controversy related to the origin of volcanic rocks and was also indirectly responsible for the economic rivalry that developed between Hartford and New Haven, which led to the construction of the failed Farmington canal. As the following accounts confirm, it also generated popular paintings and one very interesting legend.

Artistic Legacy

Odell Shepard predicted that it would be the painters, far more than the poets or prose writers, "who finally make us see, feel, and value the beauty of Connecticut and the inexhaustible variety of her moods, colors, contours, and nuances."

The production of art in colonial New England was inhibited for more than a century, first by Puritan scorn for anything ornamental and later by a lack of economic wealth among potential patrons. After the Revolution, Connecticut produced a school of itinerant painters who specialized in portraiture. Many produced twin paintings of couples with stern expressions that reflected their Puritan beliefs.

Paintings of natural scenes gradually gained access to the homes of the well-to-do and became a major art form in the 1830s and 1840s, primarily because of the success of the Hudson River School. Henry David Thoreau's writings and Thomas Cole's paintings greatly accelerated this process, which culminated in the late nineteenth and early twentieth century when American impressionists, especially those of the Lyme School, immortalized Connecticut's natural vistas. Most accomplished among the latter painters were Childe Hassam, Willard Metcalf, Alden Weir, and John Twachtman.

John Ruskin, the influential British art critic, wrote in *Modern Painters* that nature should be treated in such a way that the poet, the naturalist, and the geologist (!) should be able to derive pleasure from it. A number of artists who worked in Connecticut were influenced by this book and began to incorporate geologic scenes. Foremost among the geologic landforms that appeared in nineteenth-century landscapes are East and West Rock, the volcanic hills in the New Haven area. Benjamin Silliman, the scientist who made the earliest geologic studies of these basaltic complexes, clearly recognized their artistic prospects, describing their steep flanks as "composed of precipitous cliffs of naked frowning rock, hoary with time, moss-grown and tarnished by a superficial decomposition looking like an immense work of art."

East and West Rock are the remains of two large intrusive masses, similar in age to the Metacomet Ridge, that formed when voluminous batches of magma rose along faults and squeezed between layers of sandstone. Together with Pine Rock, Mill Rock, and Whitney Peak, those hills form a curved wall, resembling a breached medieval rampart protecting New Haven from the north.

Weathering of the iron-rich minerals in the basalt resulted in a thin veneer of orange-brown hematite (iron oxides) on its exposed surfaces. Lit by the setting sun, the steep cliffs often take on a crimson glow.

Adrian Block, the Dutch explorer, noticed this color when he sailed into New Haven's bay and referred to the rocky hills as the "Rooden Bergh," or Red Mountain.

One of the earliest artists to include a volcanic landmark in one of his paintings was Samuel Morse, who used a profile of West Rock in the background of an 1825 portrait of Benjamin Silliman. A woodcut attributed to Daniel Wadsworth and published in an issue of the *American Journal of Science* in 1830 shows the Hampden lava flow resting on a sequence of sandstone layers in the Rocky Hill Quarry. Clearly depicted is the textural difference between the subvertical cooling columns in the basalt flow and the subhorizontal layers of underlying brownstone (see fig. 5–5). Silliman's study of the baked contact between this lava flow and its underlying sediments clinched the argument among Neptunists and Plutonists about the origin of the basalt.

John Warner Barber included several volcanic masses in his work. Barber was born in Windsor. His father, a farmer of modest circumstances, died when John was only fourteen. At fifteen, Barber was apprenticed to Abner Reed, a noted engraver from whom he learned the trade. In 1834, Barber purchased a horse and buggy and traveled throughout Connecticut for three years, sketching nearly every town and village. In the winter months, he made ink wash drawings of those sketches and used them for his engravings. Barber's *Connecticut Historical Collections*, which provides snapshots of the state in the early nineteenth century, was published in 1836. Henry Howe, who became Barber's partner, wrote, "The book came upon the people like a work of magic." At the time, few citizens had traveled beyond the confines of their villages. By perusing Barber's pictures, they could imagine the world beyond and feel less isolated.

Several of Barber's prints are also of geological interest. One is a view of Meriden and two of its churches. On the right is a profile of the Hanging Hills that shows the steep western cliff face. The hill consists of a large tilted fault block composed of Holyoke basalt (fig. 5–9, top). Another print depicts a narrow defile, known as Cat Hole Pass, which separates South Mountain from Cat Hole Mountain, northeast of Meriden. Here, too, Barber emphasized the steep cliffs of the faulted Holyoke lava flow—part of the Metacomet Ridge—and its talus slopes (fig. 5–9, bottom). A particularly outstanding print represents the southeastern

Fig. 5–9. Top: Meriden, southern view of churches (John Warner Barber, 1836, page 229). Bottom: Meriden, north view of Cat Hole Pass (John Warner Barber, 1836, page 230).

view of West Rock and Westville, sketched from a location to which many artists have returned. It clearly shows the colonnaded upper section of that intrusive mass and its lower apron of basaltic rubble.

George Henry Durrie painted East Rock and West Rock many times. He was born in Hartford but spent much of his life in New Haven. Durrie began his career as a portraitist, but the winter landscapes that he painted after 1840 became his most important contribution to American art. From 1861 to 1863, Currier and Ives popularized Durrie's work by publishing ten of his paintings as hand-colored lithographs. Those images became icons of mid-nineteenth-century rural life in New England and fostered a national yearning for the simple times before industrialization. His paintings of East Rock in 1847, 1853, 1857, and 1862 show the southwest flank with its characteristic columnar cliff face and talus slope (plate 10, top). When painting West Rock in 1853 and 1857, Durrie also chose to depict the steep, columnar cliff and blocky apron at its feet, contrasting the monument's roughness with tranquil meadows, a stream, and a far-off church spire (plate 10, bottom). Durrie also favored scenes with asymmetric mountains for background, as in *Mountain Scene* (1850), *Gathering Wood* (1860), and *Cider Making* (1863); several of these imaginary mountains recall the eastward-dipping lava flows of the Metacomet Ridge.

Renowned Connecticut painter Frederic Church was also born in Hartford, to a wealthy family. When he was barely eighteen, he became an apprentice of Thomas Cole, who was nationally recognized as the country's preeminent wilderness painter. A major work that gained Church much acclaim was the painting of West Rock he finished in 1849, showing a dense forest separating the steep, rugged mountain from a pastoral scene with a meandering stream and farmers at the harvest. According to Mary Field, Church not only provided a beautiful scene but also chose West Rock to "illustrate" an important passage in colonial history: "That rugged pile recalls a story of trial and fortitude, courage and magnanimity, the noblest friendship, and a fearless adherence to political principles from religious motives." In this Victorian hyperbole, Field alluded to Edward Whalley and William Goffe, the judges who, after Cromwell's death, fled to the colonies in 1660. Pursued by royal agents, they hid in a cavity formed by large basaltic erratics high up on West Rock (fig. 2–6, top). Protected and fed by locals, they remained safe. The cave became renowned for its inscription: "Opposition to tyrants is obedience to God."

The Curse of the Black Dog

"Countless years have elapsed since the great tide of molten lava rolled over the region. Years, fewer, but still countless, have passed during which the shattered and tilted remains of the lava sheet have watched over the land. Deep gorges divide the masses into separate mountains, lonely and desolate." William Pynchon wrote those lines after visiting the Hanging Hills, part of the Metacomet Ridge north of Meriden; he used this area for a fictional encounter with a short-haired black dog. One day, while sampling basalt from outcrops in one of the gorges, Pynchon noticed a black dog trotting up the path toward him. The animal appeared friendly but kept its distance while following Pynchon throughout the day. Where Pynchon stopped to collect samples, the dog scoured the nearby wood, poking his nose into every hole and behind every stump. At the end of the day, when they had returned to the spot where they had met, the dog started off, looked back at him, and then vanished into the woods.

One evening three years later, Pynchon and Herbert Marshall, a geologist who had climbed to the west peak many times, were discussing the local geology when the subject turned to the black dog. Marshall mentioned that he had seen the same dog twice before and that the animal was the subject of a local saying: "If a man shall meet the Black Dog once it shall be for joy; and if twice, it shall be for sorrow; and the third time he shall die." Marshall laughed and said that he did not believe in omens, unless they were lucky ones. The men agreed to climb the Hanging Hills the next day.

The early spring morning was sunny but very cold. Instead of following the snow-clogged path in the Merimere Valley, they decided to ascend the steep southern face of the mountain using one of the clefts where a fault intersects the lava flow. The surface was bare except for the patches of snow and ice in hollows. When they reached the top, a cold wind blew so fiercely that they had to hold onto the ledge for support, and they quickly decided to turn back. Marshall led the way. Halfway down the slope, he abruptly stopped, turned, and pointed to the top of the cliff. High up on the rocks stood a black dog. They saw his breath steaming from his jaws, but no sound came through the biting air. Marshall's face turned white; and as he steadied himself against the rock face, he whispered, " . . . it's the third time." Even as he spoke, the ledge on which he stood gave way. There was a cry, a rattle of stones falling—and Pynchon stood alone. He quickly made his way down, but when he reached the spot where Marshall lay, bruised and bleeding, he realized that his friend had died. Later that day, men from neighboring farms climbed up to collect Marshall's body. When they approached the spot, they noticed a black dog watching over the geologist. As they got nearer, the dog swiftly fled back into the shadows of the ravine.[5]

5. On the geological map of Connecticut, a little black dog marks the basalt of Meriden's Hanging Hills. It is a warning to all geologists who climb these steep peaks and do not believe in omens.

George Edward Candee was born in New Haven and also appears to have been interested in the basalt masses surrounding the city. It is not known with whom he studied in New Haven during the 1850s. Both Nathaniel Jocelyn and George Durrie were working there around that time, and they might have influenced Candee's career. Candee painted watercolors of West Rock around 1870 and 1880. In the latter period, he also made two watercolors of East Rock. One of his works is an oil painting that shows the Sleeping Giant in the background. This intrusive mass of basalt is the same age as East and West Rock; rather than following steeply inclined fractures intersecting sandstone layers, however, its magma intruded along the more or less horizontal bedding planes of the New Haven brownstone formation to form a large blister.

Interest in the basaltic masses jutting out of the valley north of New Haven persisted into the twentieth century; John Ferguson Weir painted East Rock in the fall of 1901. Weir's work strongly contrasts with the earlier, more romantic paintings, however. The forest in the foreground is scarred, the fields are overgrown, and the wooden fence is broken (plate 11, top). Despite its natural beauty, the subject exudes desolation. The farmers have gone, their descendants probably working instead in one of New Haven's many industrial complexes.

Towers on the Metacomet Ridge

To emphasize their dominance over nature, some people climb Mount Everest; others erect towers. For several Yankees, the Metacomet Ridge was not high enough, and they constructed towers on it. In 1810, Daniel Wadsworth raised the first tower built on any mountain in the United States. It was fifty-five feet high and had a hexagonal shape; its spiral stairs allowed visitors to climb to 960 feet above sea level. From its platform John Barber wrote, "The most considerable place in sight, is Hartford, where, you see, with the aid of a glass, the carriages passing at the intersection of the streets, and distinctly trace the motion and position of the vessels, as they appear, and vanish, upon the [Connecticut] river, whose broad sweeps are seen like a succession of lakes, extending through the valley. The whole of this magnificent picture, including within its vast extent, cultivated plains and rugged mountains, rivers, towns and villages, is

Plate 9. Holyoke Basalt Ridge of Lamentation Mountain, Meriden. *Courtesy Sr. Anna Marie Goetz.*

Plate 10. Top: West Rock, New Haven, by George Henry Durrie, 1853. *Courtesy The New Haven Museum and Historical Society.* Bottom: East Rock, New Haven, by George Henry Durrie, 1853. *Courtesy The New Haven Museum and Historical Society.*

Plate 11. Top: East Rock, New Haven, by John Ferguson Weir, New Haven, about 1901. *Courtesy Florence Griswold Museum, Old Lyme; Gift of the Hartford Steam Boiler Inspection and Insurance Company.* Bottom: Ithiel Town's Bridge near East Rock, New Haven, by George Henry Durrie, 1847. *Courtesy The New Haven Museum and Historical Society.*

Plate 12. Castle Craigh, Meriden Hanging Hills. One of several towers built on the Metacomet Ridge since 1810. *Photo by Jeremy Adams.*

encircled by a distant outline of blue mountains, rising in shapes of endless variety."

The tower and grounds around Wadsworth's residence were open to the public; Daniel Webster, John Trumbull, Benjamin Silliman, and Catherine Beecher were just a few of the illustrious citizens who frequented the site. The first tower stood for only thirty years; it was blown down around 1840. Wadsworth replaced it with a similar structure that was ten feet higher. In July 1864, that tower burned down. Three years later, the new owners of Talcott Mountain, Mr. Bartlett and Mr. Kellogg, built a third tower to a height of sixty feet on higher ground that was a quarter of a mile north of the old tower sites.

Samuel Clemens (Mark Twain) visited Bartlett's tower in 1877 and, after admiring the views, wrote a thank-you note: "To Mr. Bartlett, Who has robbed the historical command / Away with him to the Tower [of London]! Of all its terror—."

In 1888, Bartlett sold his property to a wealthy New Yorker. When he learned that the latter had purchased the land and tower for private use only, however, Bartlett decided to build a fourth tower on Talcott Mountain at a point near Tariffville, overlooking the meandering section of the Farmington River. Erected in 1889, it stood seventy feet high and dominated the landscape.

The tower that best fits Connecticut's landscape was build by the Meriden industrialist William Hubbard late in the eighteenth century and dedicated on October 29, 1900. Castle Craig is constructed of angular blocks of basalt and sits on the edge of one of the basaltic ridges that make up the Hanging Hills of Meriden (plate 12).

Gilbert Hueblein, of brewery fame, had a 165-foot stone tower erected on the Metacomet Ridge for a summer residence in 1914. Its sweeping view stretches from the Berkshires to Long Island Sound. The tower survived the 1938 hurricane and is still open to the public. Megalomania by Hartford's millionaires did not stop there, however; more and more mega-mansions are joining the tower on Talcott's ridge crest.

6

The Moodus Noises

The Science and Lore of Connecticut Earthquakes

It isn't a groan, nor a crash, nor a roar,
But is quite as blood-curdling to hear,
And has stirred up more theories crammed with learned lore
Than you'd care to wade through in a year
—Reginald Sperry, 1884

The small village of Moodus lies halfway between Old Saybrook and Middletown, near the confluence of the Connecticut and Salmon Rivers (fig. 6–1). Its rocky hills are notorious for the loud noises and tremors heard and felt at intermittent intervals. Archaeological evidence indicates a high concentration of ancient Indian sites in this area and suggests that the area held special significance for native peoples (fig. 4–6). John William De Forest wrote that "the people who lived in this place had special access to the divine, and that this was the site where they heard the voice of the good spirit Kiehtan [Cautantouwit], but also the rage of Hobbamocko [Abbomocho]," the spirit responsible for human suffering and natural calamities. What consolation did they get from the one and what dark prophecy from the other?

Hammonasset, Nehantic, Wangunk, and Mohegan Indians frequented the site for religious and social gatherings, referring to it as Machemoodus [Mackimoodus],[1] the place of many noises. There they

1. Several different spellings exist for this name. F. G. Speck translated *Matchi mundu* as "devil" or "demon" in Mohegan speech. The Algonquian word for an earthquake was nanamkipoda.

Fig. 6–1. Mt. Tom, the site of the infamous Moodus noises, seen from the confluence of the Connecticut and Salmon Rivers, East Haddam (John Warner Barber, 1836, page 523).

held ceremonies to ask Kiehtan for bountiful harvests and to placate Hobbamocko.

The colonists who settled in the Moodus area occasionally heard noises emanating from Mount Tom, a hill in the form of a sugarloaf that rises a little more than three hundred feet above the Salmon River (see fig. 6–1). The north-northwest-trending hill is composed of metamorphic rocks (schists and gneisses) that contain sheets of igneous rocks (pegmatites). A second site, frequently mentioned as the source for the sounds, is Cave Hill, about a mile north-northeast of Mount Tom. Its name derives from a cave in the metamorphic rocks that is capped by a pegmatite and extends about sixty feet into the western flank of the hill. The cave, shaped like the bell of a trumpet, widens to the west. In this space, seismic vibrations in the rock are imparted to the air and amplified, causing them to be heard more loudly in the Connecticut River Valley.

In the late eighteenth century, Richard Alsop, who became known as one of the "Hartford Wits," predicted that Moodus—and its noises—would become illustrious because it was "a place celebrated for a kind of home-made earthquake, which will probably in some future day make a conspicuous figure in the natural history of the country." His two-centuries-old prediction has turned out to be quite accurate. Many different hypotheses and theories have been forwarded since to explain the Moodus noises and quakes. They reflect the early religious and later scientific beliefs about the origin of earthquakes that evolved in the emerging and maturing United States and mirror hypotheses developed in Europe.

Early Beliefs

Less than a decade after the Pilgrims landed at Plymouth, they found that their new world harbored some nasty surprises. On June 1, 1638, one day after Thomas Hooker's sermon on the Constitution, New England's ground shook violently and remained restless thereafter for several months. In a catalogue of the earthquakes that occurred between 1638 and 1869, historian William Brigham referred to the following account by an unknown author: "This year (1638) [there] was a great and fearful earthquake . . . it came with a rumbling noyse or low murmure like unto remoate thunder . . . as the noyse aproached nerer the earth begane to shake and came at length with that violence as caused platters dishes and such like things . . . to clatter and fall downe; yea persons were afraid of the houses themselves." In those days, hostile Indians, religious heretics, and poor soils threatened communities, and many felt that God sent earthquakes as well to test their religious beliefs. Rather than putting their flocks at ease, Puritan ministers made the situation worse. Peter Bulkly, pastor of Concord, warned, "It shakes, pressed with the heavy Guilt of Men; the Earth can't bear the Burden of our Sin."

During the remainder of the seventeenth century, the Massachusetts Bay region continued to experience relatively frequent seismicity. A quake on February 5, 1663, was especially strong, and its aftershocks lasted for more than a year.[2] According to a report by Samuel Williams, "The quake arrived with a sudden roar (like that of a great fire), buildings were shaken with amazing violence, walls split asunder and bells

2. The sources for the 1638 and 1663 earthquakes are unknown. The first is believed to have occurred in central New Hampshire (fig. 6–2), the second in the Malbaie region of Quebec. Both were strong, with estimated magnitudes between 6.5 and 7.

rang without being touched. Most kinds of animals sent forth fearful cries and howlings." The aftershocks kept the Massachusetts Bay Colony population on edge, especially because the tremors coincided with a period in which many social tendencies disruptive to Puritan traditions were manifesting themselves, causing concern among the clergy. The exhortations and threats from the pulpits increased.

The earliest settlers arrived in the Haddam territory east of the Connecticut River around 1655. Soon after, they began to hear the noises and experience the trembling. When the settlers inquired about the origin of the noises, an astute Wangunk apparently replied that Hobbamocko was angry because the Owanux (Englishmen) had brought their god into his territory.

In 1692, religious intolerance and persecution of those suspected of devil worship climaxed with the hanging of "convicted" witches in Salem, Massachusetts. Although that village became infamous for its witchcraft trials, Connecticut was the real witchcraft center of New England. Fifty persons were formally accused between 1647 and 1663. The majority hailed from the river towns of Hartford, Wethersfield, and Windsor. Most were acquitted, but eight women and two men, husbands of the condemned, were executed.

During that period, stories about witches began to circulate among the people of Haddam. Author William Clemmons wrote, "At Cove [Cave] Hill, where the Moodus Noises are most terrifying, the Haddam witches who practice black magic, battle with the East Haddam [Moodus] witches, dabblers in white magic, within a subterranean cave . . . while Hobbamocko umpires." When this spirit had had enough of their quarrels, he would chase them out, and "forth into the air rushed the witches with shrieks of baffled rage, and deafening peals of thunder reverberated through the hills."

The dichotomy of good East Haddam versus bad (West) Haddam witches appears to have risen from religious and political discord. East Haddam ceded from Haddam around 1700 because of "troubles" over seating the Reverend Jeremiah Hobart as minister. At the Connecticut General Assembly in May 1702, Hobart remarked, "He was praying that measures be taken to enforce his claim against the people of Haddam for arrears of salary," and expressed the hope "that the strange bellowing noises and earthquakes lately heard may awaken them to righteousness."

Despite the temporary intrusions of angry Indian spirits and shrieking witches among the locals' beliefs, the majority of Moodus's inhabitants continued to attribute the noises and quakes to the mighty hand of the Lord, who "made the hills tremble before him." In a letter written to Thomas Prince in Boston on August 13, 1729, Stephen Hosmer, the first pastor of the First Congregational Church in East Haddam, reported:

> As to the earthquakes I have something considerable and awful to tell you. Earthquakes have been here (and nowhere but in this precinct, as can be discerned; that is, they seem to have their center, rise, and origin among us), as has been observed for more than thirty years. I have been informed that in this place, before the English settlements, there were great numbers of Indian inhabitants, and that it was a place of extraordinary Indian Powwows, or in short, that it was a place where the Indians drove a prodigious trade at worshipping the devil. . . . Now, whether there be anything diabolical in these things, I know not; but this I know, that God Almighty is to be seen and trembled at, in what has been often heard among us.

Despite the popularity of religious explanations, pseudo-scientific beliefs began to surface as well. Possibly one of the strangest concerned a "book-learned and erudite man from England" by the name of Dr. Steele, who arrived in the Moodus area sometime between 1760 and 1770. In a letter to Professor Benjamin Silliman of Yale, the Reverend Henry Chapman wrote about this "scientist," who came from Europe in search of "fossils."

Steele boarded with the Knowlton family and ventured into the Moodus hills in search of carbuncles—round, white objects resembling stones in the light but remarkably luminous in the dark. According to town gossip, he took possession of an abandoned blacksmith shop in the hills, where he set up his chemical laboratory and pursued the occult (see poem by John Brainard on page 153). Soon Steele announced that he had discovered and removed the source of the noises, a large carbuncle that had outgrown its subterranean space. He predicted that there would be no more noises and quakes in the following ten to fifteen years but warned that he had seen smaller carbuncles and that these would grow to maturity and cause renewed seismic activity. One night he packed his carbuncle in lead and departed for Europe, never to be heard from again.

Packing the carbuncle in sheet lead, which was probably obtained from the lead mine in Middletown, was apparently necessary to protect against its strong effulgence, which implies a surprisingly early awareness of radioactivity. Pegmatites in the Haddam area do contain uranium-rich minerals, but Henry Becquerel, the French physicist, did not chance upon the phenomenon of radioactive decay until 1896, half a century after this legend appeared in writing, suggesting a later embellishment to the story.

Early Observations and Hypotheses

In his 1729 letter to Reverend Prince, Hosmer also described the Moodus noises:

> Whether it be fire or air distressed in the subterraneous caverns of the earth, cannot be known; for there is no eruption, no explosion perceptible, but by sounds and tremors, which sometimes are very fearful and dreadful, I have myself heard eight or ten sounds successively, and imitating small arms, in a space of five minutes. I have, I suppose, heard several hundreds of them within twenty years; some more, some less terrible. Sometimes we have heard them almost every day, and great numbers of them in the space of a year. Often times I have observed them to be coming down from the north, imitating slow thunder, until the sound came near or right under, and then there seemed to be a breaking like the noise of a cannon shot, or severe thunder, which shakes the houses, and all that is in them. They have in a manner ceased, since the great earthquake.

The "great" earthquake that Hosmer referred to probably originated off Cape Ann, northeast of Boston, on November 9, 1727, and was sufficiently strong to be felt in the Moodus region (fig. 6–2).[3] Reverend Matthias Plant described the effects of this quake on Newbury, Massachusetts:

> October 29, 1727 [Gregorian calendar] being the Lord's-Day, about forty minutes past Ten the same Evening there came a great rumbling Noise, but before the Noise was heard, or Shock perceived, our Bricks upon the Hearth

3. Relatively strong aftershocks followed in November 1727, January and February 1728, December 1730, November 1734, and August 1739.

> rose up about three quarters of a Foot, and seem'd to fall down and loll the other way, which was in half a Minute attended with the Noise or Burst. The Tops of our chimneys and Stone fences, were thrown down; and in some Places the earth opened' and threw out some Hundred loads of Earth, of a different Colour from that near the Surface . . . and in many Places, opened dry Land into Springs, which remain to this Day; and dried up Springs, which never came again. It continued roaring, bursting and shocking our Houses all that Night.

The Reverend John Lowell, another Newbury minister, wrote about "an offensive stench [sulfurous gases]; more nauseous than a putrefying corpse," that accompanied the fissuring of the land. To many, the fumes spoke of the underworld's opening and release of condemned souls. The quake drove people from their beds into the streets, where they gathered in terror-stricken groups. The next morning, churches overflowed, and large numbers of new converts were admitted in the following days. Reverend Henry White remarked "that many who had very little sense of religion before, appeared to be very serious and devout penitents."

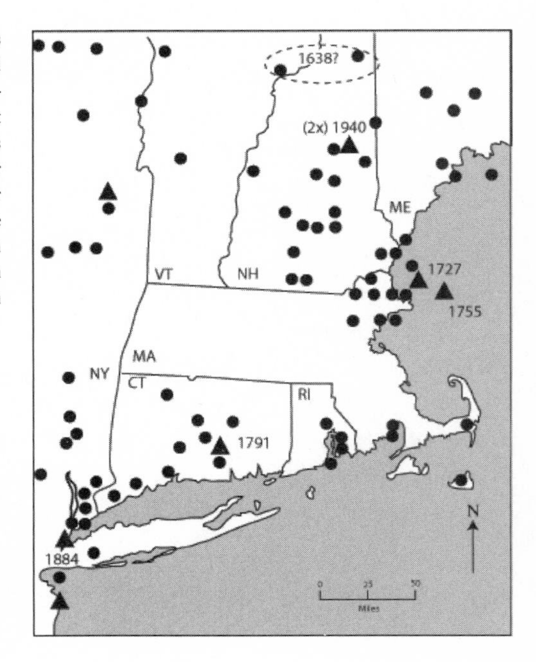

Fig. 6–2. Earthquake epicenters of western New England and New York. Dots represent earthquakes with intensities of about VI. The triangles show locations of quakes with intensities greater than VI and magnitudes greater than 5. The exact location of the 1638 quake is not known with sufficient accuracy (adapted from Weston Geophysical Research Inc., Boston).

The problems the early settlers faced had mostly receded by 1727. Agriculture was stable, commerce thrived, and the Indian threats had largely disappeared. However, community relations to God were under more than traditional suspicion. Social and economic development had brought many "sins" to the fore that tended to be accentuated in the increasingly urbanized environments. Throughout New England, the strong 1727 earthquake and its aftershocks put a temporary stop to those "sins."

Throughout the early eighteenth century, nature in the Massachusetts Bay area remained seismically restless. More than fifty minor quakes struck the Newbury area between 1727 and 1741. Strong seismic activity returned on November 18, 1755, with a major earthquake that also originated off Cape Ann and caused damage in Boston (fig. 6–2). People from Nova Scotia to Delaware felt this quake, and a Dr. Holmes, quoted by James White, provided the following description:

> Upon the first shock of the earthquake many persons jumped out of their beds, and ran immediately into the streets, while others sprung to the windows, trembling, and seeing their neighbors as it were naked, shrieked with apprehension of its being the day of judgment, and some thought they heard the last trumpet sounding, and cried for mercy . . . others expected instantly to be swallowed up and buried in the ruins . . . the brute creatures [cattle] lowed and ran to the barns for protection; the dogs howled at their master's doors; the birds fluttered in the air . . . all the animal creation were filled with terror.

The outspoken Boston ministers Jonathan Mayhew, Charles Chauncy, and Thomas Prince used the event and the following aftershocks to emphasize God's power and displeasure with the city's "sinful" inhabitants. Those Harvard-educated men, who knew the classics by heart, appear to have ignored Seneca's letter (about 62 CE) to his friend Lucilius in *Natural Questions* (book 6:3) in which he wrote: "That it would help to keep in mind that Gods cause none of these things [earthquakes] and that neither heaven nor earth is overturned by the wrath of divinities. These phenomena have causes of their own; they do not rage or command but are disturbed by certain defects, just as our bodies are." (Seneca frequently favored the analogy between the Earth and the human body.) Thirteen sermons, five poems, and three scientific pieces rapidly flowed from the printing presses of Boston. Most made it to Hartford, and Haddam's clergy and educated citizens undoubtedly read several of them as well.

In 1755, John Winthrop IV, astronomer and mathematician at Harvard College, linked the origin of New England earthquakes to the causes of volcanoes He believed both events to result from the mixture and violent reaction of various liquids and vapors that run through numerous underground caverns. These processes were seen as a natural part of the operations of a benevolent deity. He wrote that earthquakes should not be considered "absolutely evil": "It is in the physical sense alone that I say, the disjoining the parts of the earth, and opening its pores, may be the end primarily, aimed at in earthquakes, as such mutations in the earth may from time to time become necessary to the production of subterranean bodies [ores]. This view ought to silence all complaints of sufferers." That statement must surely have jolted his God-fearing colleagues at Harvard.

In 1785, Samuel Williams found it "a pretty certain fact that the earthquakes of New-England have been caused by something which has moved along under the surface of the country. What thus moved under, and hove up the surface of the earth, was probably a strong elastic vapour." Williams based his theory on the fact that the vapors had been at work several days before they became "ripe" for explosion, because wells and springs "were uncommonly altered in their motion, colour, smell and quality" up to three or four days before the 1755 earthquake struck. Such phenomena are relatively common precursors to seismic disturbances affecting swampy areas and have been used elsewhere to predict imminent seismicity.

Despite Winthrop's and Williams's attempts to introduce scientific observations into discussions about the causes of earthquakes, the clergy and general public remained convinced throughout the rest of the eighteenth century that earthquakes were the Lord's way of warning against deviations from Protestant lifestyles.

The 1755 shock was felt over a region of 300,000 square miles that included the Moodus area. Its pastors must have felt the vibrations and without doubt had copies of the earthquake sermons published in Boston. Curiously, no remaining records indicate that the people of Moodus considered this quake to be more significant than their other "noises."

The reaction in Hartford, however, was similar to that in Boston. Reverend Eliphalet Williams preached a sermon on November 23 titled "The Duty of a People, under Dark Provisions, or Symptoms of Approaching Evil, to Prepare to Meet Their God." Williams urged his flock to pray with "Uncommon Fervour, Frequency and Solemnity" and concluded, "It is of the Lord (praise ye Him) that we were not all swallowed up the other night, and instantly plunged into shoreless eternity." This is clearly a reference to what he had read about the earthquake that destroyed Lisbon, Portugal, on November 1, a little more than two weeks before the Cape Anne quake. Lisbon's quay sank into the Tagus River and disappeared below the waves, inspiring stories that the entire city had been swallowed by the earth.

The Search for Natural Causes

Connecticut experiences earthquakes more frequently than is commonly believed. Since 1678 there have been at least 137 shocks that, although weak, were felt throughout the region (fig. 6–3). A relatively strong earthquake with an epicenter in the Moodus area shook southern New England on May 16, 1791 (fig. 6–4). It arrived around 8 p.m. with two shocks in quick succession. Thirty light shocks followed almost immediately and upward of one hundred tremors were counted in the course of the night and the following morning. The *Middletown Gazette* of May 21, 1791, reported that the first shock was severe and lasted about twelve seconds, while the second was more moderate and of shorter duration.

The *Connecticut Courant* of May 23 reported a rattling, rumbling noise in Hartford, followed by a jarring shock that lasted eight to ten seconds. Near Branford, about twenty-five miles southwest of Moodus, a Captain Benedict reported schools of fish leaping in every direction out of the water in the harbor as far as his eyes could see, indicating a strong seismic wave that had compressed their air bladders. People felt the jolt in Boston and New York, which are both about one hundred miles from the source (see fig. 6–4).

Seismicity in 1791 continued with relatively strong aftershocks on May 17 and 19, followed in turn by tremors in August and October

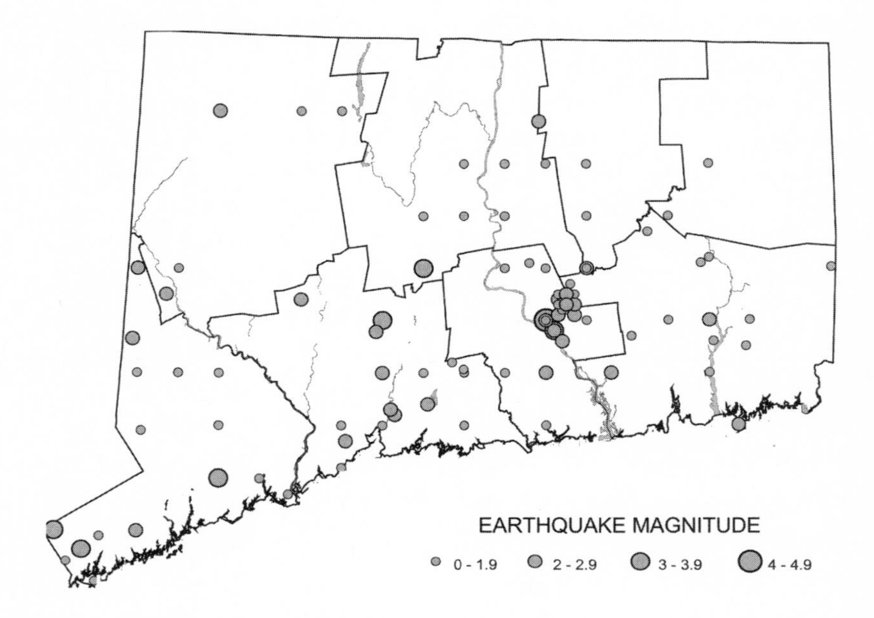

Fig. 6–3. Earthquake epicenters for 137 shocks that originated in Connecticut since 1678. Note high concentration in the Moodus seismogenic area. *Courtesy Dina Smith, Weston Observatory of Boston College.*

1792, January and July 1793, and four shocks in March 1794. Decades then passed without any appreciable activity.

The unusual occurrence of the 1791 Moodus quake and the broad region in which it was felt triggered much scientific discourse, and many hypotheses concerning its origin emerged in the following century. A diary owned by Reverend Elijah Parsons, sent to Benjamin Silliman at Yale in 1815, mentioned that the quake had shattered many chimneys, collapsed stone walls, moved boulders, and opened fissures in the ground. A chimney that stood on bedrock near the Moodus River falls was cleaved on one side from top to bottom: "Consternation and dread filled every house." Fifty years after that event, Professor John Johnson of Wesleyan University scouted the Moodus area and found no evidence of open fissures, displaced boulders, or damaged chimneys and wondered if the account in the diary could have been exaggerated. However, it seems quite possible that in the space of half a century the fissures would have filled, the boulders would have become overgrown, and the damaged chimneys would have been rebuilt.

In 1836, John Barber, a well-known historian and artist, suggested that the Moodus quakes and noises were due to "certain mineral and/or chemical combinations that caused explosions at depths of many thousands of feet below the Earth's surface." Reports provided by chemists who experimented with reactive sulfur/iron mixtures most likely influenced supporters of this hypothesis. Robert Bakewell (1768–1843), one of the earliest teachers of general geology in Britain, described such an experiment. Water, twenty-five pounds of sulfur, and an equal weight of iron filings were mixed into a paste and buried in an iron pot. A few hours later, the ground above the pot swelled and cracked. Noisy explosions of hot sulfurous vapors followed. Save for the vapors, the Moodus activity would have fitted right in with these reactions.

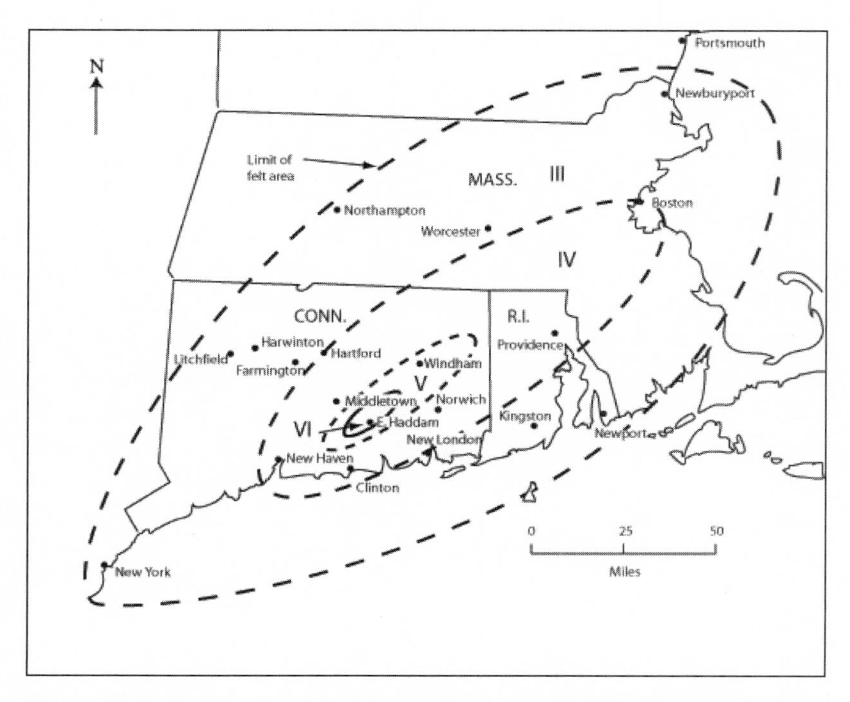

Fig. 6–4. Region in which the principal 1791 earthquake was felt. Zones represent seismic intensities from III to VI and magnitudes of about 3 to 5.

In an anonymous 1841 article on the Moodus noises in *The Classics*, a group of Wesleyan professors disagreed with Barber: "It is hardly reasonable to suppose that any disturbing cause proceeding from them [explosions of mineral compounds] should cause a sensible agitation so far distant as New York, Northampton and Boston." They proposed instead that interruptions in the natural flow of electricity in the Earth's crust caused the shocks: "the fluid [electricity] pervades all terrestrial bodies, though frequently in a latent state, and . . . may be called into activity by mechanical power, by chemical action, and by heat." They believed this naturally occurring agent to be "capable of producing all the devastations witnessed in connection with earthquakes," concluding, "We have beneath us an energy fully adequate to all those displays of power and grandeur which we witness in common electricity [lightning] around, and over our heads."

Hypotheses attributing earthquakes to electric discharges had actually emerged a century earlier. In 1756, Benjamin Franklin wrote, "If a non-electric cloud discharges its contents upon any part of the earth, when in a highly electrified state, an earthquake must necessarily ensue." The popular Reverend Thomas Prince wrote that the numerous lightning rods that appeared on Boston rooftops had "stimulated" the 1755 earthquake. Because Boston was considered the center of sin in New England and had the largest number of such rods, it had been "shaken more dreadfully." Prince held Benjamin Franklin indirectly responsible for the retribution, because he had been instrumental in developing the lightning rod.

In 1884, Emory Johnson reported that some people believed a subterranean passage led from a large cave below Mount Tom to Long Island Sound and that "certain delicate combinations of wind and tide [in this passage] were responsible for the Moodus Noises." C. F. Price concurred: "Since nature abhorred a vacuum, there must be some vast vacuum beneath Moodus' sacred hill toward which the forces of nature so violently rushed as to cause the mysterious noises." Richard Alsop satirized this new theory:

> Why Moodus groans in such convulsive frolics,
> And why Hull's physic cures all sorts of colics;
> Relinquish then the unavailing strife,
> For while I've matter left, or breath, or life,
> I'll prove . . . that Plenum *Vacuums* everywhere exist . . .

Several millennia earlier, Aristotle had attributed seismicity and subterranean noises to the effect of winds, deprived of channels through which to pass inside the Earth. He wrote, "We must suppose that the wind in the earth has effects similar to those of the wind in our bodies whose force when it is pent up inside us can cause tremors and throbbing."

In 1874, William Brigham catalogued the earthquakes that struck New England between 1638 and 1869, referring to them as "Volcanic Manifestations": "The Moodus hill has had its volcanic nature affirmed and denied . . . but the whole valley of the Connecticut River, seamed with dikes and dotted with eruptive cones, is unstudied yet. No doubt exists that volcanic agencies have been at work here in comparatively recent times." He suggested that the intrusion of basalt dikes in Connecticut was continuing underground to the present, and that "the high temperature they bring in contact with the cold rock through which they break, or into whose cavities they run, produces the tremors and disturbances we call earthquakes."

James Dwight Dana, a geology professor at Yale, strongly disagreed with Brigham's hypothesis and pointed out that the volcanic dikes and flows in the Central Valley were of Mesozoic age and several hundred million years old. In 1871, he wrote, "An earthquake is the jar from sudden fracture or displacement [in the Earth's crust]; it matters nothing whether igneous injections follow or not."

Scientific Hypotheses

In 1760, John Mitchell, a British engineer and lecturer at Cambridge University, published the most significant of the early works on earthquakes. He suggested that shifting masses of rock, many miles below the Earth's surface, caused quakes. Thirty-three years later, in 1793, Benjamin Franklin, among the leading scientists of his day, provided a mechanism for Mitchell's shifting masses: "I therefore imagined that the internal part [of Earth] might be a fluid more dense, and of greater specific gravity than any of the solids we are acquainted with, which therefore might swim in or upon that fluid. Thus the surface of the globe would be a shell, capable of being broken and disordered by the violent movements of the fluid on which it rested." Franklin's hypothesis appears to be an early precursor to the plate tectonic theory, which revolutionized geology 150 years later.

In the second part of *Faust*, written between 1827 and 1832, Johann Wolfgang von Goethe went a step further, relating seismicity to the formation of mountains (presently accepted as a geologic axiom):

> Seismos declared: "But for my batter and my clatter
> How would this world be now so fair
> How would your mountains stand above there
> In clear and splendid ether-blue
> If them I had not worked to shove there?

Almost a century after Mitchell and Franklin formulated their hypotheses, John Milne, who in 1886 compiled a detailed catalogue of British earthquakes, concluded that "seismic waves are produced by the elastic force of rocks when they spring back from their distorted form when the ground is broken and slips either up, down, or sideways." An American geologist, Karl Gilbert, provided proof for Milne's hypothesis of a connection between fracturing and earthquakes through his detailed studies of new fault scarps formed in the Owens Valley of California during the seismic activity of 1872. He concluded that fracturing was the cause, not the effect, of earthquakes. By the turn of the century, Gilbert's theory was almost universally accepted, especially after the 1906 San Francisco earthquake, but the origin of the crustal forces responsible for ruptures along faults remained in question until the plate tectonic theory was developed in the 1960s.

However, seismic activity along America's East Coast continued to baffle scientists, who could find no evidence of faults that had broken the surface during seismicity. The principal reason for the scarcity of young fault scarps in New England is twofold. First, numerous faults exist in this region, which has been tortured by plate collisions for more than half a billion years. This implies that single faults are unable to develop over long distances, and ruptures probably jump from one pre-existing fracture to another. The release of energy is therefore spread over an area, rather than in a linear (California-type) zone. Second, debris left by the last glacial event covers much of Connecticut and is especially thick in the valleys below which active faults are commonly hidden.

Evidence that Connecticut earthquakes are due to ruptures along faults has actually been available since 1855. That year, John Johnston of Wesleyan College published a report titled "Notice of some Spontaneous Movements occasionally observed in the Sandstone Strata in one of the Quarries at Portland, Connecticut." Those quarries are located about ten miles northwest of Moodus. To detach large blocks of sandstone, the quarrymen cut channels a foot or more wide and from 50 to 150 feet long into thick sandstone beds. When they had sunk an east-west channel to within nine to twelve inches from the top of the underlying layer, the remaining section would suddenly break "with a loud report by an enormous pressure, and the men in alarm leaped from the excavation they had made." Johnston reported that the northern sections of the bedrock had, on several occasions, advanced almost an inch southward by slip along the underlying bedding planes.

Referring to the deformations that took place in the Portland quarries, Edward Hitchcock, a professor of theology and geology at Amherst College, suggested at the 1863 meeting of the American Academy of Arts and Sciences that the development of such fractures "might account for certain peculiar rumbling sounds in the earth heard at times in Connecticut." He was clearly referring to the Moodus noises.

William Niles observed similar events in a quarry near Monson, Massachusetts. After a visit on June 20, 1873, he wrote, "When the opportunity is presented [during quarrying] the compressed [gneiss] rock expands with great energy, often bending, and fracturing the beds, and sometimes producing sudden and violent explosions, rending and displacing the rocks and occasionally throwing stones and other debris into the air." Loud noises heard many miles away accompanied the deformation. A few days after such an event, he found the place "looking much as though a small but powerful earthquake had taken place." A five-foot-thick layer of gneiss had ruptured along two parallel fissures about eighty feet long. Niles estimated the lateral shortening at about 1.5 inches. As Hitchcock had before him, Niles attributed the "strange sounds in the earth, so frequently and candidly reported" in Connecticut, to a process of deformation resembling that which had occurred in the quarries. He concluded, "May not the same power produce some of the slight earthquake shocks in non-volcanic districts?"

Responsible Forces

In spite of all the theorizing, the origin of the powerful, unfathomable forces inside the earth remained a mystery. Wilbur Foye, a Wesleyan professor, believed that the force responsible for the Moodus quakes resulted primarily from readjustments following the depression of Connecticut's rocky crust by the weight of land ice during glacial periods. His 1949 hypothesis could in fact explain the stress releases that had occurred in the Portland and Monson quarries. In Connecticut, the land ice had generally moved from the north-northwest to the south-southeast (fig. 2–5). A mass of ice about one mile thick is equivalent in weight to a layer of rock 1,800 feet thick. Thrusting such a layer southward would have stressed the upper segment of the underlying bedrock and left it compressed. The Earth's crust remembers such deformation for tens of thousands of years. The slow process of unloading after the ice melted slowly raised the land and intermittently released the "fossil" strain (the degree of deformation left in the rock), leading to upward and southward shifts of its rock masses close to the surface.

Fig. 6–5. Offset drill holes along Route 2 in the Salem area, indicating recent southward thrusting caused by stress that was probably introduced by moving land ice during the last glacial advance.

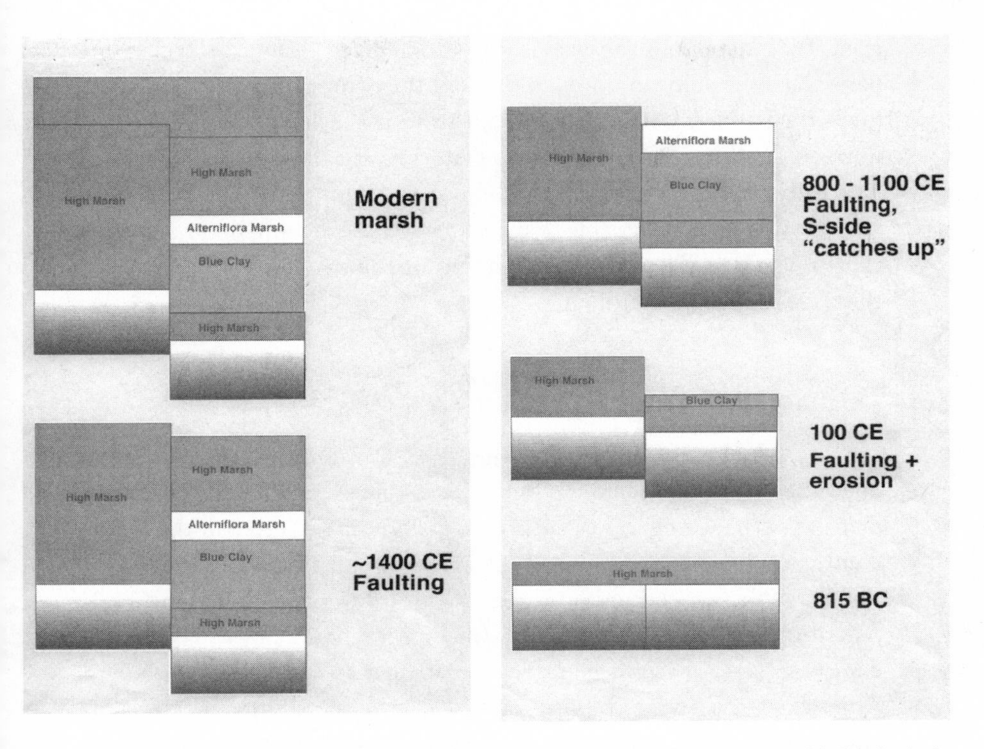

Fig.6–6. Offset sediment layers on either side of a major fault. Reconstructed from a series of drill holes in the Farm River marsh near Branford. *Courtesy Johan Varekamp, Wesleyan University.*

Evidence supporting Foye's hypothesis was encountered twenty miles east of Moodus during construction of Route 11. Holes left by drilling deep into the Hebron gneiss were separated and offset by up-dip shifts of the upper segments of the outcrop along low-angle northerly inclined fault planes (fig. 6–5). The thrusting was uniformly south-southeast, parallel to the direction in which the former ice sheets that covered the region had moved. Here, as in several quarries, the stress release was associated with a manmade disturbance of the bedrock.

Sediments in a coastal salt marsh near Branford, twenty-five miles southwest of Moodus, provide the most recent evidence for post-glacial tectonic activity. In 1999, William Thompson, a Wesleyan graduate student, reported significant vertical differences of up to 3.5 feet between sediment layers of similar age on either side of a buried east-northeast trending fault (fig. 6–6). Rupture appears to have occurred in several stages over a period of at least 2,800 years. The most recent event that caused a recognizable offset took place around 1400 CE. Slippage was caused by an up-dip motion of the bedrock north of the fault. Such thrusting most likely resulted from a south-southeast-directed horizontal compression introduced during the last glacial event. A relatively strong tremor occurred in this region in 1858, and in the last twenty years, seismographs have registered seven small earthquakes that appear associated with further rupture along this fault, indicating that stress releases continue in the area.

Deformation of the North American Plate below Moodus

In the 1960s, geologists, aided by new discoveries in geophysics, developed the concept of plate tectonics. This theory holds that the movements of crustal segments, or plates, over a ductile layer in the upper mantle are the principal causes of fracturing and associated seismicity. For example, spreading and the growth of new crust along the Mid-Atlantic Ridge, which represents the eastern boundary of the North American plate, force the continent westward. In California, the North American plate's western border collides with, and slides along, the Pacific plate. Both zones are characterized by frequent seismicity. There was no reason therefore to believe that forces at work in those faraway tectonic zones could affect regions in the central part of the North American plate. However, the Appalachians are seismically active, although earthquakes occur less frequently and generally are less intense. Recent discoveries suggest that two opposing forces control the motion of this plate: the growing Mid-Atlantic Ridge, which pushes the continent westward, and a viscous drag along the plate's bottom that opposes this motion. The latter is especially effective in regions with deep crustal roots, such as the Appalachians. The deeper this keel, the greater the resistance as the plate floats westward.

Interest in Moodus's seismicity peaked in the 1960s, during construction planning for the Yankee Power Plant on the Connecticut River. To make sure that this structure could resist seismic activity, historical earthquake records were analyzed. N. H. Heck believed that the principal 1791 Moodus quake had a Modified Mercali intensity of VIII (8) on a scale with twelve steps. Daniel Linehan, a seismologist from Weston Observatory in Boston, disagreed. He estimated the maximal intensity of the shaking at VI (6) or possibly V (5). His much lower assessment greatly benefited the electric company, and the plant was built to withstand intensity V (5) shaking. The disparity was surprising because the two seismologists had used the same historical evidence.

A decade after construction of the plant, seismologists from the National Oceanic and Atmospheric Administration reviewed the available information again and recommended that the 1791 quake be reclassified as intensity VII (7), which equals a Richter magnitude of about 4.35.[4] Earthquakes with magnitudes of 4 or less generally do little damage, especially to structures built in and on bedrock. However, because different parts of a complex nuclear plant resonate at different frequencies, even a minor quake could have resulted in major problems.[5]

In 1974, recurring seismicity in the Moodus area motivated the Nuclear Regulatory Commission to fund a seismic network around the plant. Five stations were in place by 1979, in time to record two intense swarms of micro-earthquakes—more than one thousand events affected the area from August through mid-October 1981 and in June 1982. Additional seismic swarms occurred in the late 1980s (1986 and 1987–88). The most detailed information was obtained from a temporary array of seismometers deployed in the region of the 1987 activity. The data suggested ruptures at relatively shallow depth along one or more northeast-trending faults. Locally, activity might have jumped from one such fault to an adjacent one via pre-existing northwest fractures.[6]

To obtain more information about the forces responsible for the fracturing and seismic activity, three holes were drilled to depths of 900, 1,500, and 4,800 feet respectively. Down-hole tests and field data indicated the presence of a north-northwest/south-southeast

4. Globally, more than six thousand tremors with magnitudes between 4 and 4.9 occur each year; the number of smaller tremors (M = 3–3.9) has been estimated at around fifty thousand. The amount of energy released by an M4 quake is about thirty times greater than that of an M3 tremor. Tremors with magnitudes below 2 are "felt" only by sensitive instruments.

5. The nuclear plant was recently decommissioned, and all that is left on the site is a flat, grassy area. Forty casks with more then one thousand highly radioactive rods, however, are stored in a valley adjacent to the former site.

6. Earthquakes with magnitudes greater than 2 but less than 3 have been rather common in the Moodus area. John Ebel of Weston Observatory and Boston College found evidence for seven such events between 1913 and 1925 and two each in 1939, 1951, 1968, and 1976. There were seven quakes with this magnitude in the 1980s.

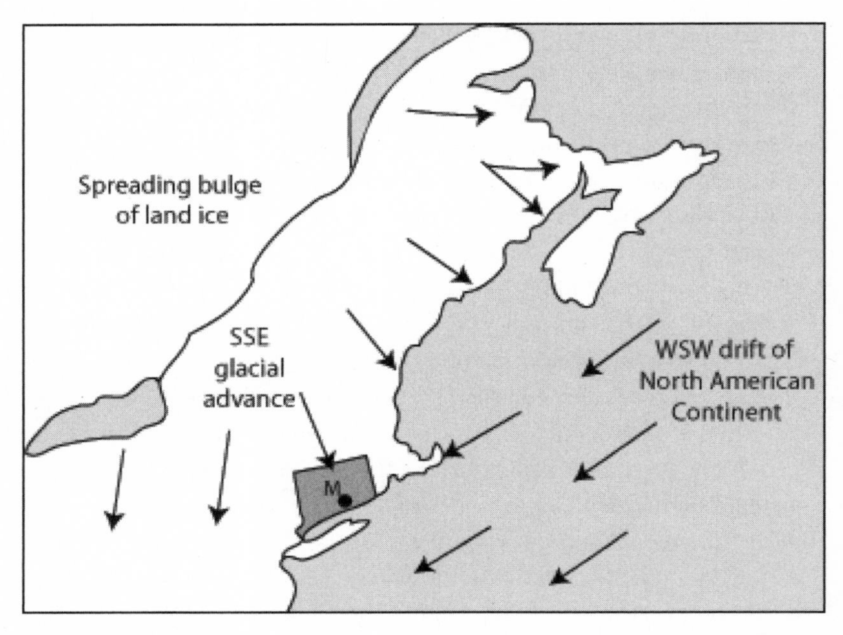

Fig. 6–7. Directions of horizontal forces responsible for the seismicity in New England. Shallow (to a depth of less than 800 feet) south- and southeast-directed stresses were probably caused by previous loading and movement of ice masses during the latest glacial event. The growth of new crust in the spreading Atlantic basin results in a west-southwest stress that forces the North American continent westward and dominates at deeper levels.

directed (horizontal) compressive stress at shallow levels and an east-northeast/west-southwest (horizontal) compressive stress below 800 feet in the deep hole.[7]

The drill data thus indicate that two different compressive stress configurations might affect Moodus. The smaller south-southeast-directed stress at shallow levels was most likely introduced during the most recent glacial event. Its fossil strain should gradually decrease over time. The much stronger west-southwest-directed stress that dominates at levels below 800 feet is most likely related to the drift of the North American plate (fig. 6–7). This force has affected

7. The west-southwest-directed horizontal compressive stress increased from 40 mPa (mPascal, a unit used to measure tectonic stress) at 800 feet to 90 mPa at 3,800 feet, in Moodus's deep hole. The latter value is sufficiently high to cause slippage along pre-existing fractures, and to generate earthquakes.

the continent, and thus Connecticut, since the opening of the Atlantic Ocean basin and will continue to cause seismicity in the weak crustal zones of eastern North America.

Some seismologists believe that the return period for New England quakes with an intensity of VI (6) is several decades, those with intensity of VII (7), several centuries, possibly longer. Sensitive seismographs have provided better estimates of the seismic activity and cumulative stress releases in the last half of the twentieth century, but that period is too short for predicting believable recurrence intervals.

As the Mid-Atlantic Ridge continues to push North America westward, no doubt exists that earthquakes will visit Connecticut in the future. What should one do when scientists will ultimately be able to predict quakes? Mark Twain's advice: "[When] you make up your mind that the earthquake is due, you stand from under, and take hold of something to steady yourself, and the first thing you know you get struck by lightning."

Moodus Poetry

In 1825, John Brainard, editor of the *Hartford Mirror*, published the poem "Matchit Moodus," which relates the story of Dr. Steele, the alchemist who discovered the carbuncles he thought responsible for Moodus's earthquakes and noises.

> O'er Moodus river a light has glanced,
> On Moodus hills it shone;
> On the granite rocks the rays have danced,
> And upwards those creeping lights advanced
> Till they met on the highest stone.
> By that unearthly light, I see
> A figure strange, alone,
> With magic circlet on his knee,
> And, decked with Satan's symbols, he
> Seeks for the hidden stone.
> Now upward goes that gray old man,
> With mattock, bar and spade—

The summit is gained and the toil begun,
And deep by the rock where the wild lights run,
The magic trench is made.
Loud, and yet louder was the groan,
That sounded wide and far;
And deep and hollow was the moan
That rolled around the bedded stone
Where the workman plied his bar.

In 1884, Reginald Sperry published "Moodus Noises, a rhyme for the fourth of July." The poem starts with a description of the village's reputation as a place of noises and then relates the noisy prank that mischievous Simeon Strong, a boy of sixteen—"a real Yankee" known throughout the county as "Shankey"—plays on the inhabitants.

In a certain respectable state "way down east"
Is a village of title euphonious;
Though the village itself doesn't differ so much
From others familiar in story—
Yet by one startling feature 'tis quite widely known,
Which rescues its name from obscurity;
This curious thing that confers such renown
Is, if you'll believe me, a noise
Which, coming at intervals, floats o'er the town
Like Damocles' sword held in poise.
It isn't a groan, nor a crash, nor a roar,
But is quite as blood-curdling to hear;
And has stirred up more theories crammed with learned lore
Than you'd care to wade through in a year.

Shankey secretly stashed a load of fireworks at the mouth of the grotto at Cave Hill. To celebrate Independence Day, the citizens of Moodus gathered near the river to watch a display of lights, listen to the brass band, and get "plumb-loaded" in various ways. When the festivities were in full swing, Simeon snuck through the woods to the cave and ignited the fireworks. The result was a general panic (fig. 6–8):

Fig. 6–8. East Haddam revelers running for their lives after Shankey and friends ignited some fireworks at the mouth of the Moodus cave during the Fourth of July festivities along the river (illustration accompanying *Moodus Noises, A Rhyme for the Fourth of July*, by Sperry).

At the first hiss the crowd had all paused, and now stood
In silence, as though transfixed by magic;
From their looks you'd have said they were figures of wood
With faces most woefully tragic
No wonder—for look! From the cave comes a flash
Of fires—green, red, blue and yellow—
Then their hearts almost stop as they hear a great crash
Followed close by an unearthly bellow.
Then comes an explosion—the earth fairly cracks—
And a line of hot something comes streaming—
Human nature can't stand it—some fall on their backs,
And some fall praying—some screaming
One man makes a dash for the river and dives,
While others go climbing the beeches;
But most of them turn, and just flee for their lives,
Making everything blue with their screeches.

Moodus Tremors and Sonic Booms

In the early morning of June 17, 1981, while enjoying a bowl of cornflakes, I heard a low rumbling sound. Shortly thereafter, the windows and sliding doors of our house in Haddam rattled strongly. Our three dogs panicked and joined me en masse on the couch. We had experienced a small earthquake, a Moodus tremor. After wiping spilled milk, cornflakes, and dogs from the couch, I walked (fast) down our path to the Haddam Meadows, anxious to see whether there was any unusual activity at the Yankee Power Plant across the river. All was quiet on the nuclear front!

That afternoon two similar events occurred. One appeared to last a little longer and turned out to have been two tremors in rapid succession. A seismic instrument buried in our backyard recorded the vibrations and sent its signals to Weston Observatory in Boston for analysis. It turned out that the tremors belonged to a pulsating swarm of more than one hundred seismic events spread over a period of two months. The sensitive seismic network picked up most of the tremors. The strongest shock was felt throughout the Haddam–Moodus area and had a magnitude estimated at 2.9.

After that event I became more attuned to possible Moodus tremors and "discovered" many very faint ones that rattled the sliding doors around the same time, between eight and nine in the morning. Weston Observatory was silent and the dogs slumbered on. Finally I found out that the Concorde leaving Kennedy airport would break through the sound barrier somewhere high above Long Island and cause a shock wave that at times made it all the way to the surface. I should have paid more attention to the dogs!

Seismologists in the Netherlands have told me that their instruments have registered tremors with magnitudes up to 1.3 caused by crowds dancing during rock concerts. That's one way to experience a typical Moodus "quake"!

Visitors from Space

The Weston and Wethersfield Meteorites

The question is not what God's got against Wethersfield,
it's what God's got for Wethersfield.
—Barbara Narendra, 1982

Most people have looked up and seen meteors flash through the night skies. Their lights flare and vanish within seconds. Unknown to them, some of these objects continue their voyage and reach the Earth's surface, occasionally even invading homes.

After he woke up early in the morning of April 8, 1971, Paul Cassarino walked downstairs and noticed a fine coat of whitish dust covering the floor and furniture of his living room. Looking up at the ceiling, he saw a substantial dent, surrounded by an array of small cracks. During the night, a meteorite weighing about 12.3 ounces had broken into his Wethersfield home without waking him or his wife.[1] The black stone had crashed through the roof and lodged in the ceiling of the living room.

Early in the evening of November 8, 1982, a little more than a decade after Cassarino's foreign visitor arrived, Mr. and Mrs. Robert Donahue of Wethersfield were watching TV when they heard what appeared to be "the muffled roar of a Mack truck coming through the front door." They leaped up and ran into the living room, which they found in chaotic disorder: dust, wood splinters, and pieces of plaster

1. *Meteor* is used for the streak of light produced in the upper atmosphere by the entry of a small body from space, *meteorite* for an object that survived its passage through the atmosphere and landed on the Earth's surface. Asteroids are small planet-like bodies (or fragments thereof) that orbit the sun.

Fig. 7–1. The 1982 meteorite that crashed through the roof of the Donahue home in Wethersfield, and the hole it left in the ceiling. *Courtesy Dan Haar, Hartford Courant.*

covered the floor and furniture. When they looked up, they noticed a sizeable hole in the ceiling. The air was filled with what they thought was smoke. Realizing that something had crashed into the house, they went outside and saw a large gap in the roof just above the front door. Firefighters called in to assess the damage found a crusty black rock (fig. 7–1) weighing almost six pounds below the table in the dining room. The meteorite had blown a hole through the roof and ripped through a closet on the second floor before entering the living room (fig. 7–1, bottom). From there, it jumped into the adjacent dining room and bounced around until it wound up under the table.

Within a relatively short period, two meteorites had hit homes less than two miles apart in the same residential town south of Hartford. The chances of this happening defy all odds. In an article in the *Wethersfield Post,* the events were hyperbolically referred to as "a fantastic solar coincidence, unmatched in history." Douglas Preston wrote that the probability of two meteorites falling on houses in the same town approximated "the chance of two blind flies colliding somewhere in the Grand Canyon."

When a TV reporter asked the rhetorical question "What does the man upstairs have against Wethersfield?" Barbara Narendra, from Yale's Peabody Museum meteorite division, replied, "The question is not what God's got against Wethersfield; it's what God's got for Wethersfield." Obviously, some unique presents!

Meteorites have hit homes elsewhere in the United States. There are about twenty authenticated reports of such collisions in the twentieth century, two of which are especially interesting. On November 30, 1954, two pieces of a meteorite fell two miles apart in Sylacauga, Alabama. The larger one, weighing 8.5 pounds, crashed through the roof of the Hodges home at about 1:00 p.m. It penetrated the ceiling, bounced around, and headed for the couch, where Mrs. Hodges lay fast asleep. She was struck on the hip and hand and suffered serious bruises.

Another meteorite came very close to Connecticut, but it landed in Peekskill, New York, in 1992. The brilliant string of fireballs containing the stone was seen over part of the eastern seaboard, traveling northeast (fig. 7–2). Sonic booms were heard as the meteorite passed over North Carolina and Pennsylvania. Michelle Knapp heard a loud crash around 8:00 p.m. and went outside to check on its cause. At first, she saw nothing extraordinary. Then she noticed damage to her 1980 Chevy Malibu, which was parked in front of the house. A twenty-six-pound stone had crashed into its trunk, bending and twisting the metal. The Peekskill stone is one of the few meteorites for which an orbit could be calculated with reasonable precision. Part of its oval track coincided with the inner edge of the asteroid belt, suggesting that it had probably originated in that region of the solar system.

As with the path of the Peekskill stone, several observers witnessed the fiery track of the Donahue meteor that landed in Wethersfield.[2] In

2. Friction caused by high-speed flight through the atmosphere heats the outer layer of the meteor, continuously stripping molten material from it and producing a trail of incandescent gas. At heights between six and ten miles above the Earth's surface, the frictional melting ends, quickly extinguishing the fireball.

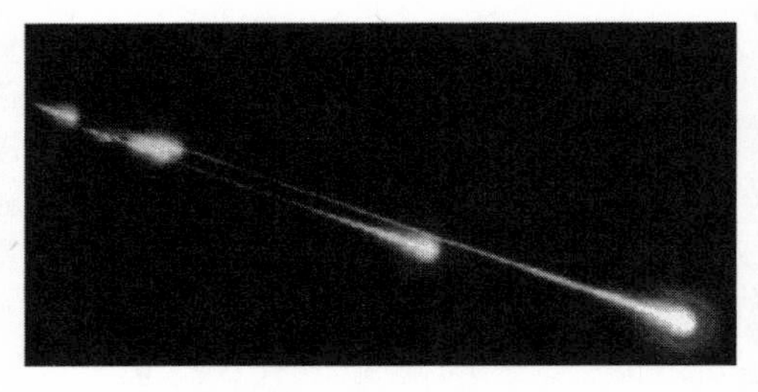

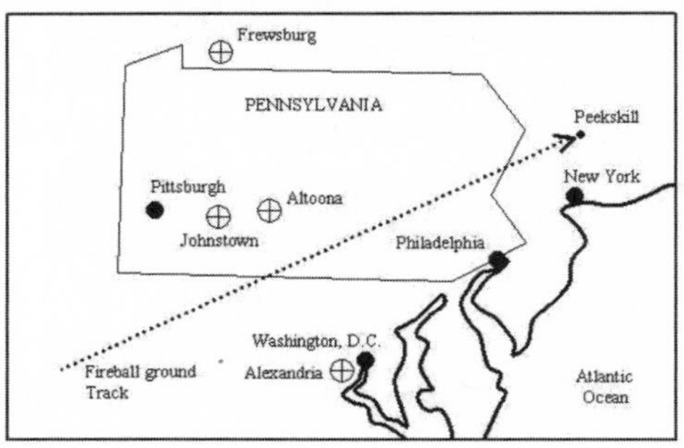

Fig. 7–2. Top: Peekskill meteor traveling east-northeast in 1992, showing it breaking up above Altoona, Pennsylvania. Photo by S. Eichmiller. Bottom: Flight path of the Peekskill meteor and observation sites.

an interview Ted Pace described driving near Mahwah, New Jersey, about eighty-four miles southwest of Wethersfield, when he noticed an object streaking across the sky. "It was plunging down in a fiery trail and had a reddish-orange color. . . . Pieces came off it." Robert de Collibus of Marlborough, Massachusetts, about seventy miles east-northeast of Wethersfield, described the entry as "bright as the full moon . . . very white." He mentioned a yellowish tail that lingered in the sky for several minutes. Stan Hedden, a resident of Glastonbury, across the river from Wethersfield, saw "a flash like lightning, illuminating the entire sky." He heard a series of six or more "rifle shots" coming from the direction of Wethersfield. Hedden's observation suggests that the original mass had broken up into several fragments during its passage through the atmosphere.

Many Wethersfield residents went looking for other pieces but found none. This should not have come as a surprise. The meteorite shower that fell west of Homestead, Iowa, in 1875 covered an elliptical area of about twenty square miles. More than one hundred stones were eventually found, those at opposite ends of the elliptical zone were as much as six miles apart. Based on the locations of the holes made in the roof and ceiling, it could be ascertained that the first Wethersfield meteorite had come from the northeast and the second from the west-northwest. Those looking for more specimens of the later fall should search the Glastonbury Meadows and the hills east of the Connecticut River.

Meteoroids enter the Earth's atmosphere at speeds between eight and forty-five miles per second; the slowest meteor could travel from New York to Los Angeles in five minutes. Differences in their flight paths cause this great variation in velocity. Some objects meet Earth, which moves at eighteen miles per second, head on; others have to catch up to it and therefore enter its atmosphere at lower speeds. The ones with the slower entries are the most likely to survive, although they can lose more than 90 percent of their initial mass during heating and ablation in the atmosphere. Small stones that enter at speeds greater than twenty miles per second usually disintegrate.

An earlier unexpected solar visitor surprised Connecticut's citizens in 1807. A swarm of meteorites hit the southwestern part of the state on December 14, at about half past six in the morning. The meteor came from the north and entered at a shallow angle. Nathan Wheeler of Weston, one of the justices of the Court of Common Pleas for Fairfield County, described the principal fireball as having a diameter about half or two-thirds that of a full moon. As the large globe of fire raced south, it flashed with a vivid light, dulling somewhat but remaining clearly visible behind stretches of thin clouds. Its smoky tail faded rapidly in the morning light. Wheeler estimated the entire period between its first appearance on the northern horizon and the total extinction of its tail at about thirty seconds. Seconds after the fireball extinguished, witnesses heard three loud and distinct explosions, indicating that the meteor had broken up. A rapid succession of smaller overlapping blasts followed, resembling the rumbling sound associated with thunder. Meteorite fragments were found at six sites in a north-south belt stretching over a distance of about ten miles from Huntington to eastern Easton (fig. 7–3).[3]

Benjamin Silliman and James Kingsley, professors at Yale College, heard about the fall and immediately set out to collect samples and interview witnesses. Merwin Burr of Huntington had been standing on

3. The event became known as the Weston meteorite fall, but its fragments actually fell mostly in the Easton area, which later seceded from Weston and was incorporated in 1845.

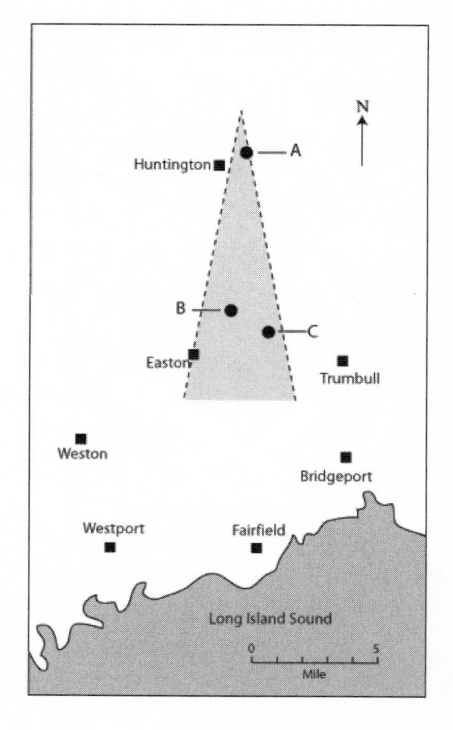

Fig. 7–3. Map of possible strewn field of the "Weston" 1807 meteorite between Huntington and Easton, Connecticut. A: Merwin Burr residence, Huntington; B: William Prince residence, Easton (formerly Weston); C: Tashua Hill, Trumbull.

the road in front of his house when the most northern of the stones landed about fifty feet from him. It hit a rocky outcrop, and the impact reduced much of the meteorite to powder and small fragments. The largest piece, still warm to the touch, was smaller than a goose egg. From the remains, Silliman and Kingsley estimated that the original stone weighed about twenty to twenty-five pounds before impact.

The next find was made five miles south of Burr's house. William Prince and his family were still in bed when they heard a loud noise, followed by an explosion. Looking through a window, Prince noticed a fresh hole dug into the turf of his front yard, about twenty-five feet from the house. Strangely, he paid no further attention to it at the time. Later that day, though, he investigated the hole and found a meteorite of about thirty-five pounds at a depth of two feet. The blackened rock had yellow and silver specks on fresh breaks. Noticing those minerals, Prince assumed that it contained gold and silver and proceeded to hammer it into small pieces. The largest piece that survived his prospecting weighed about twelve pounds.

Silliman and Kingsley reported that the Weston meteorites contained small concentrations of brilliant gold pyrite, known as fool's gold. They also contained shiny whitish specks of nickel-iron, which could easily be mistaken for silver. They wrote, "Strongly impressed with the idea that these stones contained gold and silver, they [some of the owners] subjected them to all the tortures of ancient alchemy, and the goldsmith's crucible, the forge, and the blacksmith's anvil, were employed in vain to elicit riches which existed only in the imagination."[4]

Six days later, more stones were discovered half a mile northwest of Prince's house. The meteorite had shattered when it landed on an outcrop of gneiss and had probably weighed between seven and ten pounds before impact. Half a mile northeast of Prince's property, another meteorite that originally weighed about thirteen pounds had broken into two pieces on impact. A fifth stone fell two miles southeast of Prince's home, at the foot of Tashua (Tashowa) Hill. Ephraim Porter reported seeing smoke rising from the ground. At the time, he supposed that lightning had struck, but after hearing about the rain of stones, he went looking and found a number of meteorite fragments inside and around a hole in the ground that was about twenty inches in diameter and two feet deep. Silliman and Kingsley estimated the weight of that meteorite to have been twenty to twenty-five pounds.

The largest single mass fell in a field belonging to Elijah Seely, about 165 yards from his home. Elihu Staples, who lived on the hill above the field, witnessed its fall. After the explosion, there was a riveting noise like that of a whirlwind, and a streak of light passed over Staples's orchard. Then "a shock was felt and a report heard like that of a heavy body falling to the earth." When Seely went out to take care of his cows, he found a hole about five feet in diameter. Schist fragments and clumps of grassy soil had been thrown a distance of one hundred feet. The meteorite had broken up on impact. The largest piece was found on Tashua Hill, a few days after Silliman and Kingsley had scoured the area. It was sold to Colonel George Gibbs, a noted mineral collector, whose collection was bought by Yale University eighteen years later. That stone weighed 36.5 pounds (fig. 7–4).

4. Ironically, meteorites have become very valuable these days and have appeared at Sotheby's auctions. A Sikhote-Alin iron meteorite that fell in Russia on February 12,1947, brought $122,750 in October 2007.

Fig. 7–4. The largest (recovered) meteorite that fell in the Weston/Easton area (originally 36.5 pounds, now about 28 pounds). *Courtesy Barbara Narendra, Peabody Museum of Natural History, New Haven.*

Silliman and Kingsley estimated the total weight of all stones recovered in the area to have been about 330 pounds.

In 1811, Nathaniel Bowditch, an astronomer, published estimates of the height, direction, and velocity of Weston's meteoroid, using observations by William Page of Rutland, Vermont; Mrs. Gardner of Wenham, Massachusetts; and Judge Wheeler of Weston. After a careful series of calculations, Bowditch concluded that the meteor had traveled in a southerly direction, more exactly, seven degrees west of due south (S 7° W) and nearly parallel to Earth's surface, at a height of about eighteen miles before breaking up. He estimated its velocity to have been more than three miles per second and believed that the greatest part of the original mass had traveled farther south and fallen into Long Island Sound or beyond.

Reports and newspaper accounts dealing with the Weston meteorite were received with much skepticism. President Thomas Jefferson read several of the stories and allegedly remarked, "It's easier to believe that the two Yankee professors would lie than that stones should fall from heaven." Most Americans shared the sentiment. Scientists across the Atlantic, however, showed great interest. Published accounts were read aloud during meetings of the Philosophical Society in London and the Académie des Sciences in Paris.

With his careful analysis of the Weston meteorite, Silliman put himself in the early ranks of American scientists who believed that "rocks could indeed fall from heaven." He gave public lectures on meteorites in New York, Boston, Hartford, and other cities, complete with props. A specially constructed compartment in his carriage allowed him to travel with about fifty pounds of different kinds of meteorites.

Silliman predicted that the impacts of large meteorites or comets could be catastrophic. Around 1840, he wrote, "May they not one day come down entirely? Shall we desire it? They might sweep away cities and mountains; deeply scar the earth and rear from their own ruins, colossal monuments of the great catastrophe." He clearly foresaw the discovery of large impact sites a century later. It was not until the 1940s that the cosmic origin of the meteor crater in Arizona was accepted by the scientific community.

Of the approximately 330 pounds of the Weston meteorite that probably fell, less than fifty pounds can be presently accounted for. Collectors and scholars spread its fragments across the globe. In addition to the museums of several major U.S. cities, pieces are housed in Berlin, Calcutta, Mainz, Moscow, Munich, Ottawa, Paris, Vienna, Rome, and the Vatican.

Scientists equipped with increasingly refined research tools continue to study the fragments today. Two dozen publications with new data on the texture, chemistry, and time the Weston meteorite spent in space have appeared since 1965. The studies show that its principal minerals are olivine, orthopyroxene, plagioclase, nickel-iron, and chromite, a combination of minerals believed to be typical for matter in Earth's mantle. In atomic percentages, the common Earth elements of silica, magnesium, and iron are most abundant, making up more than 90 percent of the rock.

Wethersfield and Weston were not the only towns in Connecticut where stones fell from heaven. On May 27, 1974, around four in the afternoon, a single walnut-sized stony meteorite weighing less than two ounces landed in the city of Stratford. It made a small dent in the asphalt of the street. Normally, it would have been ignored, but a "whistling" sound accompanied its fall and led to its retrieval.

Numerous Visitors: From Dust to Danger

People have observed meteorite falls from time immemorial. For thousands of years, they believed that the objects were celestial messages sent by the gods. Some interpreted meteorites as good omens; others thought they presaged evil. Much probably depended on where they fell and how big they were.

Tens of thousands of tons of extraterrestrial debris shower Earth annually, but most of it arrives as a fine dust. One of the greatest observed showers occurred on the night of November 12, 1833. Professor Denison Olmsted of Yale described a continuous succession of fireballs that resembled rockets radiating in all directions from a single point in the heavens. The meteors were of various sizes and degrees of splendor; some were mere points, but others were larger and brighter then Jupiter and Venus, and one appeared nearly as large as the moon. The flashes of light were so bright that they woke people in their beds.

Occasionally, larger fragments survive their passage through the atmosphere. On the evening of June 24, 1938, the northern part of Pittsburgh experienced a brilliant light and was rocked by a tremendous explosion. It turned out to be a stony meteorite of several hundred tons that disintegrated into a cloud of dust barely twelve miles above the surface. Had it remained intact a little longer, it could have wiped out part of the city.

Most meteorites dive into the oceans that cover more than two-thirds of the Earth's surface. Of the remainder, many fall unnoticed in deserts, tundras, and jungles.

On rare occasions, however, large masses smash into the surface, leaving sizeable craters. About two hundred sites where large asteroids and comets have collided with the Earth are presently known by their astroblemes, the "star wounds" they left behind on land. Their craters vary enormously in size. The Haviland (Kansas) Crater is only thirty-three feet in diameter, while the largest surviving impact site in North America, the Sudbury astrobleme of Ontario, has a diameter of about 125 miles, more than twice the size of Connecticut. The most widely known is the Barringer Crater, also referred to as the Meteor Crater, in Arizona, which is three-quarters of a mile across and more than five hundred feet deep. It was caused by a mass of nickel-iron weighing more than fifty thousand tons that struck Earth about fifty thousand years ago.

Origin and Chemistry of Meteorites

The meteorites that fell in Connecticut are chondrites, the most common type of cosmic stone. Chondrites contain chondrules, small spherical particles that can comprise as much as three-quarters of the meteorite. The chondrules formed when molten, or partially molten, droplets of hot, silicate-rich liquid cooled during condensation of a primordial cloud more than 4.5 billion years ago.

Their texture and composition provide a record of the earliest events in the history of our solar system. A chondrite's chemistry is very close to the type of solar matter that remains when the sun's gaseous constituents, hydrogen and helium, are removed. Such common elements as oxygen, silicon, iron, and magnesium make up more than 90 percent of a chondrite's total weight. Most minerals that formed in chondrites, such as olivine, feldspar, and pyroxene, are similar to those found on Earth.

Based on their iron content, three chemically distinct groups of ordinary chondrites have been distinguished: high iron (H), low iron (L), and low total iron/low metalliferous iron (LL). The Weston meteorite has been classified as a high iron (H4) chondrite, the Wethersfield and Stratford stones as low iron (L6) chondrites.

Carbonaceous chondrites make up an additional group but have not been found in Connecticut. They are rare and especially interesting. As their name suggests, they are relatively rich in complex carbon compounds and oxygen. Some contain water-bearing minerals such as serpentine and gypsum. Carbonaceous chondrites represent some of the matter from which Earth could have gained the specific ingredients essential for the origin of life.

In addition to chondrites, there are achondrites, stony meteorites without chondrules, and iron meteorites. The former resemble the volcanic/igneous rocks of Earth and its moon; the latter are composed mainly of nickel-iron, the type of matter that may be present in the Earth's core. Achondrites could be parts of the outer crusts of some large asteroids, and the irons could be parts of their cores, before these bodies were shattered during collisions.

Chondrites do not fit such hypothetical models, however. Their sun-like chemistries and ages suggest that they are the earliest matter that formed in Earth's solar system, and that achondrites and nickel-iron meteorites might derive from chondritic material that accumulated and differentiated, forming the cores, mantles, and crusts of planets shattered during collisions.

Scientists use meteorites to estimate the age of the Earth and so far have dated more than seventy of them. Their ages indicate that the solar system, and hence the Earth, formed sometime between 4.53 and 4.58 billion years ago.

Meteorites and Impact Craters

The meteorites that have fallen in Connecticut are small. The largest single stony meteorite found in the United States weighed 2,360 pounds and fell on February 18, 1948, near Norton, Kansas. The largest American iron meteorite was discovered in 1902 near Willamette, Oregon. It is a deeply scarred, conical mass of nickel-iron weighing 15.5 tons. Robert Peary, the polar explorer, found a large mass of this type in Greenland—Ahnighito is eleven feet long and weighs thirty-four tons. It took three Arctic seasons to drag this heavy black mass from its bed to the shore of Cape York. Once on board Peary's ship, the meteorite's strong magnetism caused much confusion, because it put all compasses out of order. Thirty horses were needed to haul Ahnighito through the streets of New York to its new resting place in the American Museum of Natural History.

No geological evidence exists for major impacts in Connecticut. This does not mean that such an event could not have occurred, because long intervals of time are missing from the state's geological record. The oldest rocks exposed in the state formed about one billion years ago, which is less than 22 percent of the Earth's age. The possibility that Connecticut was missed by a major asteroid, however, does not mean that it was not indirectly affected by collisions elsewhere in the world.

Impacts closest to Connecticut occurred in and offshore from present-day Virginia and Delaware. One of the impact sites is located on the continental shelf off Atlantic City; the other lies beneath the entrance to the Chesapeake Bay. The latter crater is about twenty-five miles in diameter and surrounded by a highly fractured zone of rocks for a total diameter of about fifty miles, but it is hidden below a thick blanket of younger sediments (fig. 7–5). The responsible asteroids formed part of a swarm that streaked in from the north and collided

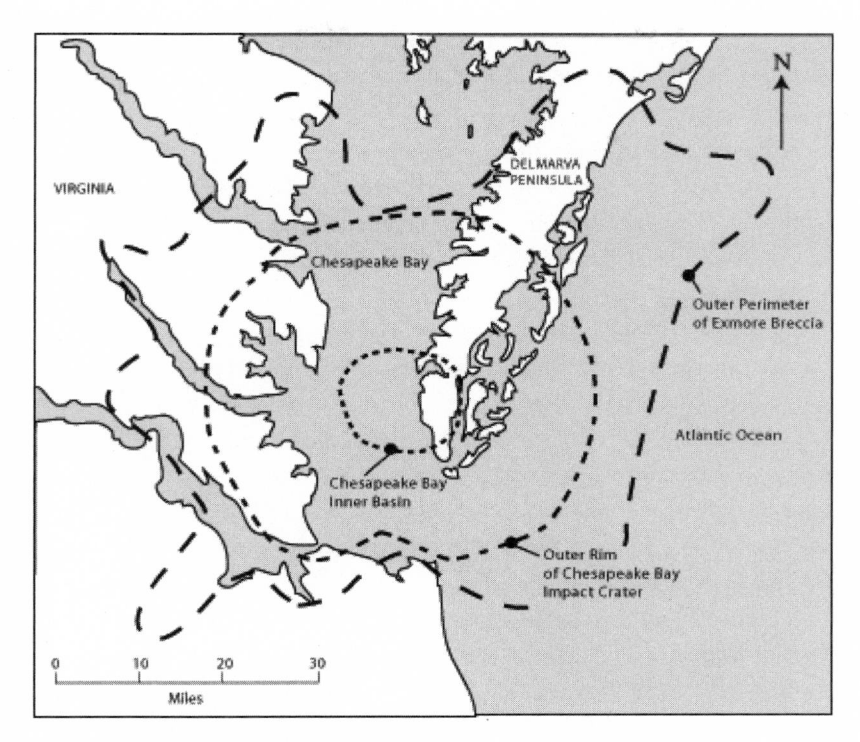

Fig. 7–5. The Chesapeake Bay impact site showing inner basin and outer rim. A layer of broken-up rock fragments (breccia) extends locally more than forty-five miles from the center. Molten fragments (tektites) that shot into the atmosphere were found as much as a thousand miles away (adapted from Poag et al., 1994, figure 1).

with Earth about 36 million years ago. Its largest mass caused the Popigai Crater in Siberia, which has a diameter of about sixty-five miles.[5]

As these asteroids entered the atmosphere, they would have resembled objects much larger and brighter then the sun. Huge cones of superheated, compressed air, veritable blow torches, would have preceded the bodies, forcing super-heated air across the earth's surface in fiery blasts that would have instantaneously scorched every living thing for many hundreds of miles. On impact the Earth would have been jolted, as if by a very strong earthquakes, and the shock waves would have bored their way to depths of more than four thousand feet,

5. Samples collected from the Popigai Crater suggest that the asteroids were low iron, L-type chondrites, chemically similar to the ones that visited Wethersfield and Stratford.

leaving a deep hole. Dense clouds of brecciated rock fragments and droplets of molten stone would have shot back up through the atmosphere into the stratosphere.

Near the Chesapeake site, the thickness of the layer of dust and debris has been estimated at more than twenty-seven feet. What was left of Connecticut's ecology after the firestorm was covered with a layer of dust and small rock fragments about three feet deep that rapidly eroded in the following years.

The Chesapeake impact was so severe and generated so much heat locally that millions of rock fragments melted and were blasted into the atmosphere, where they cooled rapidly and formed glassy rocks, or tektites. The largest number of these stones, more than 1,700, has been found in east-central Georgia. Tektites related to this event, however, have also been encountered in seafloor sediments of the western Atlantic, the Gulf of Mexico, and the southern Caribbean, a region totaling several million square miles.

The collisions produced globe-encircling clouds of dust, dimming the sunlight for months and causing a period of climate deterioration that lasted for many years. This event might therefore have been responsible for a period during which the Earth experienced major ecological problems and transitioned from the Eocene to the Oligocene epoch.

One of the greatest worries of our age is the likelihood of a nuclear war. Largely overlooked by many, however, is the potentially more destructive threat posed by rogue asteroids and comets. There are more than two thousand asteroids with diameters bigger then a half mile near the Earth's orbital path, and comets continually enter the inner solar system. Toutatis, an irregularly shaped asteroid that is 3.75 miles long, passed the Earth at a distance of only one million miles in September 2004. The energy that would be released by the impact of such a body is far larger than that of the worst volcanic eruption, the strongest earthquake, or the most advanced cluster of nuclear bombs. While thousands of scientists are employed in the weapons industry, only a few have been assigned to watch for incoming projectiles from space. This fatalistic attitude may be understandable, however, because humanity might survive a nuclear disaster but not the impact of a major asteroid!

Afterword

Our Lithic Inheritance

The Introduction began with a narrative of a hike in one of the wooded sections of Connecticut and an encounter with a stone wall that once enclosed a farmer's field, which is presently surrounded by dense growth. Right there, a very small but essential part of Connecticut's history was revealed in its basic simplicity.

Continuing the trail uphill, one can see that fewer and fewer glacial rocks of the type used to build stone walls crop out and patches of bedrock, shaved smooth by glaciers and fringed by mosses, become more common. The path connects to a deeply rutted and partially overgrown road that leads to a small quarry. Rows of holes a few inches deep show how slabs of rock have been pried from the outcrop with hammer and chisel. Judging from the size of the quarry, one may conclude that barely enough rock was removed to construct the foundations for two or three farm houses and their barns. Similar stone walls and little quarries cover much of the state, a testimony to the extent to which colonists depended on their lithic environment. From early on, Connecticut's land and its geology had a major influence on its people, but in the last century it has become clear that the Yankees have an even greater influence on the land. In the last few decades alone, central Connecticut has gradually become part of a metropolis that stretches from Washington, D.C., to Boston. The Central Valley is clogged with clusters of homes, tall buildings, and highways. Along the coast, wetlands have been dredged to accommodate marinas and filled in to provide building lots. Only six state parks where families can go to spend a day at the beach can be found along the 240-mile coastline. Inland, rivers have been impounded, and quarries bite large gaps into ridge crests.

The present-day capacity of Connecticut's citizens to erode the land significantly exceeds that of nature's.

To save at least part of the land from further abuse, The Nature Conservancy has designated twenty-one sites as preserves and wildlife areas, and the state has set aside land for ninety parks and thirty state forests. Although some of these areas contain valuable geologic entities, their principal aim is to preserve the ecology. Only two sites exist to protect important geologic features: the Dinosaur State Park in Rocky Hill and the Brownstone Quarry in Portland. Most quarries have instead become private property, and one of the most famous, the Strickland pegmatite quarry near Portland, is now part of a golf course. Other important sites, such as Southington's Great Unconformity, Durham's fossil fish locality, and Salisbury's Bashful Lady Cave, are insufficiently protected or presently closed to the public.[1]

Geology is simply not considered to be as important as ecology, despite the fact that while most plants and animals repopulate, a crystal or fossil, once removed, can never be reborn. In 1677, Matthew Hale wrote in *The Primitive Origination of Mankind*, "Minerals are a degree below vegetables." In those days, the physical environment was often perceived as antagonistic, and rocks were the most difficult to deal with. Nowadays, caring about our ecology has become fashionable, but little appears to have changed on the geologic side. Rocks and minerals still rank far below the birds and the bees.

Children entering mineral exhibits in museums are entranced by the variety of colors and shapes and can't believe that such splendor grows underground. Their interest piqued after such a visit, they often pick up the first attractive rock they see along the road or on a beach and bring it home. In most cases, those rocks wind up in the backyard, banned from the home under the motto "rocks belong outside." Fieldstone fireplaces and granite kitchen countertops, on the other hand, are accepted because they serve a "purpose."

Such selective discrimination needs to stop! A common rock or mineral, especially a native one, should sit on everyone's mantelpiece, bookshelf,

1. Some (misplaced) protection can be found in a little park in Hartford, where the sculptor Carl Andre arranged thirty-six boulders, saved from the crusher, in a triangular space. The price of this protection came to more than $2,400 per rock, much of it provided by the National Endowment for the Arts.

or windowsill to remind the whole family of the beauty that exists in the thin crust that supports our civilization. Maybe each family should have a "pet rock." The first generation of pet rocks conquered the national scene in 1975. They came with certifications of pedigree, training manuals, and eligibility for "beauty" contests. After a short life span, they too often wound up in backyards.

Those among us who do appreciate the beauty of rocks and their minerals accumulate collections that often contain specimens that have been hacked (or blasted) out of rare Connecticut outcrops. These specimens have certainly found some form of protection, but they can be enjoyed only by their owners, and they could have taught us more about their origin and history when they were still in situ. Geotopes, geological features, and sites that provide information on the evolution, structure, and properties of the Earth's crust should be set aside throughout Connecticut. They include characteristic exposures of sedimentary, metamorphic, and igneous rocks, especially those with rare fossils or uncommon minerals. They also include outstanding topographic features such as quarries, caves, glacial moraines, and eskers. Geotopes are especially valuable for research and teaching, but they are equally important to our history. They are our lithic inheritance!

Bibliography

Introduction (pp. 1–7)

Cone, J. "The Song of the Connecticut River." *The Connecticut Magazine*, vol. 7, pp. 563–68, 1902.

Morse, Jedidiah. *The Traveller's guide, or, Pocket gazetteer of the United States: Extracted from the latest edition of Morse's Universal gazetteer.* New Haven: N. Whiting, 1832.

Shepard, O. *Connecticut: Past and Present.* New York: Alfred Knopf, 1939.

1. In the Beginning: Continental Fusion and Breakup (pp. 8–26)

PRIMARY SOURCES

Badash, L., ed. *Rutherford and Boltwood, Letters of Radioactivity.* New Haven: Yale University Press, 1969.

Barber, J. W. *Connecticut Historical Collections: History and Antiquities of Every Town in Connecticut.* New Haven: B. L. Hamless, 1836.

Becquerel, A. H. "Ubereinige Eigenschaften der X-Strahlen des Radiums." *Physikalische Zeitschrift*, vol. 6, pp. 666–69, 1905.

Boltwood, B. B. "The Disintegration Products of Uranium." *American Journal of Science*, vol. 23, pp. 77–88, 1905.

———. "On the Ultimate Disintegration Products of the Radioactive Elements: The Disintegration Products of Uranium." *American Journal of Science*, vol. 25, pp. 253–67, 1907.

Cone, J. "The Song of the Connecticut River." *The Connecticut Magazine*, vol. 7, pp. 563–68, 1902.

Deford, D., ed. *Flesh and Stone: Stony Creek and the Age of Granite.* Stony Creek, Conn.: Leetes Island Books, 2000.

Dillard, A. *Teaching a Stone to Talk: Expeditions and Encounters.* New York: Harper and Row, 1982.

Hesse, H. *Siddhartha.* Transl. Hilda Rossner. New York: Bantam Books, 1971.

Hutton, J. *Theory of the Earth; with Proofs and Illustrations.* Edinburgh: Printed for Messrs. Cadell and Davies, 1795.

King, C. "Catastrophism and Evolution." *American Naturalist*, vol. 11, no. 8, pp. 449–70, 1877.

Leclerc de Buffon, G. L. "L'histoire naturelle, générale et particulière." *Les epoques de la Nature*, vol. 20, 1778.

Milford. S. *The Complete Poetical Works of William Cowper.* London: Oxford University Press, 1913.

Percival, J. G. *Report on the Geology of the State of Connecticut.* New Haven: Osborn and Baldwin, 1842.

Playfair, J. *Hutton's Unconformity.* Transactions of the Royal Society of Edinburgh, vol. 5, no. 3, 1805.

Shepard, O. *Connecticut: Past and Present.* New York: Alfred Knopf, 1939.

Silliman, B. "Sketch of the Mineralogy of the Town of New Haven, Connecticut." *Academy of the Arts and Sciences Memoir*, vol. 1, pp. 83–96, 1810.

———. *Consistency of the Discoveries of Modern Geology with the Sacred History of the Creation and the Deluge.* New Haven: Hezekia Howe and Co., 1833.

Thompson, W. (Lord Kelvin). "On the Secular Cooling of the Earth." *Mathematical and Physical Papers*, vol. 3, pp. 295–311. London: C. J. Clay and Sons, 1890.

Ussher, James. *Annalis Veteris at Novi Testamenti.* 1658.

SECONDARY SOURCES

Badash, L. "Rutherford, Boltwood and the Age of the Earth: The Origin of Radioactive Dating Techniques." *Proceedings of the American Philosophical Society*, vol. 112, pp. 157–69, 1968.

Barrell, J. "Central Connecticut in the Geological Past." Connecticut Geological and Natural History Survey Bulletin 23, 1915.

Bell, M. "The Face of Connecticut: People, Geology and the Land." Connecticut Geological and Natural History Survey Bulletin 110, 1985.

Brown, E. B. "A Man of East and West: Geologist, Savant, and Wit." *Century Magazine*, vol. 80, p. 382, 1910.

Cook, T. A. *Geology of Connecticut.* Hartford: Bond Press, 1933.

Davis, W. M. "The Triassic Formation of Connecticut." *U.S. Geological Survey Annual Report* 18, part 2, pp. 1–192, 1898.

Foye, W. G. "The Geology of Eastern Connecticut." Connecticut Geological and Natural History Survey Bulletin 74, 1949.

Jorgensen, N. *A Guide to New England's Landscape.* Chester, Conn.: Globe Pequot Press, 1977.

King, C. "The Age of the Earth." *American Journal of Science*, vol. 45, pp. 1–20, 1893.

Little, R. *Dinosaurs, Dunes, and Drifting Continents: The Geology of the Connecticut Valley.* Easthampton, Mass.: Earth View, 2003.

Longwell, C. R., and E. S. Dana. *Walks and Rides in Central Connecticut.* New Haven: Tuttle, Morehouse, and Taylor, 1932.

McHone, Gregory. *Great Day Trips to Discover the Geology of Connecticut.* Wilton, Conn.: Perry Heights Press, 2004.

McIntrye, D. B., and A. McKirdy. *James Hutton: Founder of Modern Geology.* National Museum of Scotland, 2006.

McDonald, N. G. "The Connecticut Valley in the Age of Dinosaurs: A Guide to the Geologic Literature, 1681–1995." State Geological and Natural History Survey of Connecticut Bulletin 116, 1996.

Merril, G. P. *The First One Hundred Years of American Geology.* New Haven: Yale University Press, 1924.

Raymo, C., and M. E. Raymo. *Written in Stone: A Geological History of the Northeastern United States.* Old Saybrook, Conn.: Globe Pequot Press, 1989.

Repcheck, J. *The Man Who Found Time.* Cambridge, Mass.: Perseus, 2003.

Rice, W. N., and H. E. Gregory. "Manual of the Geology of Connecticut." Connecticut Geological and Natural History Survey Bulletin 6, 1906.

———. "Guide to the Geology of Middletown and Vicinity." Connecticut Geological and Natural History Survey Bulletin 41, 1927.

Rodgers, J. "The Geological History of Connecticut." *Discovery,* vol. 15, no. 1, pp. 3–25, 1980.

Van Dusen, A. E. *Connecticut.* New York: Random House, 1961.

Ward, J. H. *The Life and Letters of James Gates Percival.* Boston: Ticknor and Field, 1866.

Wight, C. A. "James Gates Percival." *The Connecticut Magazine,* vol. 6, pp. 87–92, 1900.

Wilson, L. G. "Benjamin Silliman's Influence on Nineteenth-Century American Science." In *Connecticut: Past, Present, and Future; Transactions of the Connecticut Academy of Arts and Sciences,* vol. 56, pp. 25–35, 1999.

Wyse Jackson, P. N. *The Chronologer's Quest: The Search for the Age of the Earth.* Cambridge: Cambridge University Press, 2006.

MAPS

Altamura, J. J. "Bedrock Mines and Quarries of Connecticut." Connecticut Geological and Natural History Survey, 1987.

Fritts, C. E. "Bedrock Geology of the Southington Quadrangle, Connecticut." U.S. Geologic Survey Quadrangle Map GQ 200 (+ text), 1963.

Radway Stone, J., J. P. Schafer, E. Haley London, M. L. DiGiacomo-Cohen, R. S. Lewis, W. B. Thompson. "Quaternary Geologic Map of Connecticut and Long Island Sound Basin." United States Geological Survey Scientific Investigations Map 2784, 2005.

Rodgers, J. "Bedrock Geological Map of Connecticut." Connecticut Geological and Natural History Survey, 1985.

2. Weather and Climate: Hurricanes and Ice Ages (pp. 27–55)

PRIMARY SOURCES

Bair, F. E., et al. *The Weather Almanac.* Detroit: Gale Research Inc., 1992.

Barber, J. W. *Connecticut Historical Collections: History and Antiquities of Every Town in Connecticut.* New Haven: B. L. Hamless, 1836.

Bell, M. "The Face of Connecticut: People, Geology, and the Land." Connecticut Geological and Natural History Survey, Bulletin 110, 1986.

Blair, T. A. *Climatology: General and Regional.* New York: Prentice Hall, 1942.

Blodget, L. *Climatology of the United States.* Philadelphia: Lippincott & Co., 1857.

Bradford, W. *History of Plymouth Plantation, 1606–1646.* New York: C. Scribner's Sons, 1908.

Brumbach, J. J. "The Climate of Connecticut." Connecticut Geological and Natural History Survey Bulletin 99, 1965.

Byron, Lord G. G. "Darkness." In *The Works of Lord Byron.* Philadelphia: Carey and Hart, vol. 4, pp. 146–48, 1843.

Cable, M. *The Blizzard of '88.* New York: Athenaeum, 1988.

Dobson, P. "Remarks on Boulders." *American Journal of Science,* vol. 10, pp. 217–18, 1826.

Dwight, T. *Travels in New England and New York.* New Haven: Published by the Author, 1821.

Egger, H., and E. Bruckl. "Gigantic Volcanic Eruptions and Climate Change in the Early Eocene." *International Journal of Earth Science,* vol. 95, pp. 1065–70, 2006.

Franklin, B. A letter to Ezra Stiles, 1763. In *The Papers of Benjamin Franklin,* ed. L. W. Labaree, vol. 10, pp. 264–67, 1959–95 series. New Haven: Yale University Press.

Grove, J. M. *The Little Ice Age.* New York: Methuen, 1988.

Hampstead, J., 1717. Mentioned in Bell, M., "The Face of Connecticut: People, Geology, and the Land." Connecticut Geological and Natural History Survey, Bulletin 110, p. 68, 1986.

Holmes, O. W. "The September Gale." *New-England Galaxy,* July 2, 1830.

Hoyt, J. B. "The Cold Summer of 1816." *Annals of the Association of American Geographers,* pp. 118–31, 1958.

Jacoby, G. C., and E. R. Cook. "Past Temperature Variations Inferred from a 400-Year Tree-Ring Chronology from Yukon Territory, Canada." *Arctic and Alpine Research,* vol. 13, no. 4, pp. 409–18, 1981.

Koteff, C., J. Radway Stone, F. D. Larsen, G. M. Ashley, J. C. Boothroy, and D. F. Dincause. "Glacial Lake Hitchcock, Postglacial Uplift and Post-lake Archaeology." Fieldtrip Guidebook. Contribution no. 63, pp. 169–78, Department of Geology and Geography, Amherst, Massachusetts, 1988.

Loomis, A. "The Meteorology of New Haven." Transactions of the Connecticut Academy of Sciences, vol. 1, 1866.

Manchester, W. R. *The Glory and the Dream.* Boston: Little, Brown and Co., 1973.

Mansfield, C. "Account of Frost, Summer, 1816." In H. Stommel and E. Stommel, *Volcano Weather,* pp. 37–38. Newport, R.I.: Seven Seas Press, 1983.

Mather, C. "Essay of Cold." In *The Christian Philosopher,* p. 81, 1721.

Mather, I. *An Essay for the Recording of Illustrious Providence: Wherein an Account Is Given of Many Remarkable and Very Memorable Events, Which Have Happened This Last Age, Especially in New England.* New York: Wiley, 1856.

McIntyre, S., and R. McKitrick. "Corrections to the Mann et al. (1998) Proxy Data Base and Northern Hemisphere Average Temperature Series." *Energy and Environment,* vol. 14, pp. 751–71, 2003.

Milankovitch, M. *Théorie mathématique des phénomènes thermiques produits par la-radiation solaire.* Published for the Serbian Academy of Sciences and Arts. Paris: Gauthiers-Villards,1920.

Olsen, P. E. "A 40-Million-Year Lake Record of Early Mesozoic Orbital Climatic Forcing." *Science,* vol. 234, pp. 842–48, 1986.

Perry, C. E. *Founders and Leaders of Connecticut, 1633–1783.* New York: Heath and Co., 1934.

Petit, J. R., J. Jouzel, D. Raynaud, N. I. Barkov, J.-M. Barnola, I. Basile, M. Bender, J. Chappellaz, M. Davis, G. Delaygue, M. Delmotte, V. M. Kotlyakov, M. Legrand, V. Y. Lipenkov, C. Lorius, L. Pépin, C. Ritz, E. Saltzman, and M. Stievenard. "Climate and Atmospheric History of the Past 420,000 Years from the Vostok Ice Core, Antarctica." *Nature,* vol. 399, pp. 429–36, 1999.

Post, J. D. *The Last Great Subsistence Crisis in the Western World.* Baltimore: John Hopkins University Press, 1977.

Redfield, W. C. "Remarks on the Prevailing Storms of the Atlantic Coast of the North American States." *American Journal of Science,* no. 20, pp. 17–51, 1831.

Ridge, J. C., and F. D. Larsen. "Re-Evaluation of Antevs' New England Varve Chronology and New Radiocarbon Dates of Sediments from Glacial Lake Hitchcock." *Geological Society of America Bulletin,* vol. 102, pp. 889–99, 1990.

Shepard, O. *Connecticut: Past and Present.* New York: Alfred Knopf, 1939.

Stiles, E. *The Literary Diary of Ezra Stiles.* Ed. F. B. Dexter. New York: Charles Scribner's Sons, 1901.

Twain, M. *The Complete Short Stories and Famous Essays of Mark Twain.* New York: Collier and Son, 1923.

Utley, V. "Hurricane." *New York Spectator,* September 30, 1815.

White, H. *The Early History of New England.* Boston: Sanborn, Carter and Bazin, 1841.

Wilder, T. *Three Plays: The Skin of Our Teeth.* New York: Harper and Row, 1942.

Williams, S. *The Natural and Civil History of Vermont.* Middlebury, Vermont: S. Mills, 1809.

Williamson, H. *Observations on the Climate in Different Parts of America.* New York: T. and J. Swords, 1811.

Zachos, J., M. Pagani, L. Sloan, E. Thomas, and K. Billups. "Trends, Rhythms, and Aberrations in Global Climate 65 Ma to Present." *Science,* vol. 292, pp. 686–93, 2001.

SECONDARY SOURCES

Allen, E. S. *A Wind to Shake the World.* Federal Writers Project (1938). Boston: New England Hurricane WPA, 1976.

Antevs, E. "The Recession of the Last Ice Sheet in New England." *American Geographical Society Research Series,* vol. 11, 1922.

Beckman, J. E., and T. J. Mahoney. "The Maunder Minimum and Climate Change." *ASP Conference Series,* vol. 153, pp. 1–6, 1998.

Black, R. F. "Modes of Deglaciation of Connecticut: A Review." In *Late Wisconsin Glaciation of New England,* ed. G. J. Larson and B. D. Stone, pp. 77–100. Dubuque, Iowa: Kendall Hunt Publishing Co., 1982.

Cronon, W. *Changes in the Land: Indians, Colonists, and the Ecology of New England.* New York: Hill and Wang, 1982.

D'Arrigo, R. D., E. R. Cook, M. E. Mann, and G. C. Jacoby. "Tree-Ring Reconstructions of Temperature and Sea-Level Pressure Variability Associated with the Warm-Season Arctic Oscillation since AD 1650." *Geophysical Research Letters*, vol. 30, no. 11, pp. 31–34, 2003.

De Menocal, P. B. "Cultural Responses to Climate Change During the Late Holocene." *Science*, vol. 292, pp. 667–72, 2001.

Donnelly, J. P., N. W. Driscoll, E. U. Chupi, L. D. Keigwin, W. C. Schwab, E. R. Thieler, and S. A. Swift. "Catastrophic Meltwater Discharge down the Hudson Valley: A Potential Trigger for the Intra-Allerod Cold Period." *Geology*, vol. 33, no. 2, pp. 89–92, 2005.

Eddy, J. A. "The Maunder Minimum: A Reappraisal." *Solar Physics*, vol. 89, pp. 195–207, 1983.

Ennis, B. P. "The Life Forms of Connecticut Plants and Their Significance in Relation to Climate." Connecticut Geological and Natural History Survey Bulletin 43, 1928.

European Project for Ice Coring in Antarctica. "Eight Glacial Cycles from an Antarctic Ice Core." *Nature* 429, pp. 623–28, 2004.

Fagan, B. *The Little Ice Age: How Climate Made History, 1300–1850*. New York: Perseus Books, 2000.

Fleming, J. R. *Historical Perspectives on Climate Change*. Oxford: Oxford University Press, 2000.

Flint, R. P. "The Glacial Geology of Connecticut." Connecticut Geological and Natural History Survey Bulletin 47, 1930.

Franklin, B. *The Papers of Benjamin Franklin*. Ed. L. W. Labaree, vol. 10, pp. 264–67. New Haven: Yale University Press, 1959–95 series.

Goodrich, S. *Ridgefield in 1800*. Hartford: Acorn Club, 1954.

Harington, C. R., ed. *The Year without a Summer? World Climate in 1816*. Ottawa: Canadian Museum of Nature, 1992.

Hays, J. D., J. Imbrie, and N. J. Shackleton. "Variations in the Earth's Orbit: Pacemaker of the Ice Ages." *Science*, vol. 194, pp. 1121–32, 1976.

John, B. S. *The Winters of the World: Earth under the Ice Ages*. New York: Wiley, 1979.

Jones, P. D., A. E. J. Ogilvie, T. P. Davies, and K. R. Briffa. *History and Climate: Memories of the Future?* New York: Kluwer Academic Publishers, 2001.

Kirk, J. M. "The Weather and Climate of Connecticut." Connecticut Geological and Natural History Survey Bulletin 61, 1939.

Koteff, C., and F. Pessl. "Systematic Ice Retreat in New England." U.S. Geological Survey Professional Paper 1179. Washington, D.C.: U.S. Government Printing Office, 1981.

Lane, F. W. *The Elements Rage*. New York: Chilton Books, 1965.

Ludlum, D. M. *Early American Hurricanes, 1492–1870*. Boston: American Meteorological Society, 1963.

———. *Early American Winters I, 1604–1820*. Boston: American Meteorological Society, 1966.

———. *Early American Winters II, 1821–1870*. Boston: American Meteorological Society, 1968.

Macdougall, D. *Frozen Earth: The Once and Future Story of Ice Ages*. Berkeley: University of California Press, 2004.

Minsinger, W. E. *The 1938 Hurricane: An Historical and Pictorial Summary.* Boston: Blue Hill Observatory, 1988.

Muller, R. A., and G. J. MacDonald. *Ice Ages and Astronomical Cause: Data, Spectral Analyses and Mechanisms.* New York: Springer Verlag, 2000.

Palley, P. A. "Climate Data Sources in Connecticut." Connecticut Agricultural Experiment Station Bulletin 461, 1981.

Perley, S. *Historic Storms of New England.* Salem, Mass.: The Salem Press, 1891.

Post, J. D. *The Last Great Subsistence Crisis in the Western World.* Baltimore: John Hopkins Press, 1977.

Rampino, M. R., and S. Self. "Climate-Volcanic Feedback and the Toba Eruption of ~74,000 Years Ago." *Quaternary Research,* vol. 40, pp. 69–80, 1994.

Redfield, J. H. *Life in the Connecticut River Valley, 1800–1840.* Essex, Conn.: Connecticut River Museum, 1988.

Ribes, J. C., and E. Nesme-Ribes. "The Solar Sunspot Cycle in the Maunder Minimum AD 1645–1715." *Astronomy and Astrophysics,* vol. 276, pp. 549–63, 1993.

Ridge, J. C., and F. D. Larsen. "Re-Evaluation of Antevs' New England Varve Chronology and New Radiocarbon Dates of Sediments from Glacial Lake Hitchcock." *Geological Society of America Bulletin* 102, vol. 7, pp. 889–99, 1990.

Scotti, R. A. *Sudden Sea: The Great Hurricane of 1938.* New York: Little, Brown & Co., 2003.

Stone, J. R., and G. M. Ashley. "Ice-Wedge Casts, Pingo Scars, and the Drainage of Glacial Lake Hitchcock." *New England Intercollegiate Geological Conference,* no. 66, pp. 305–31, 1992.

Tannehill, I. R. "Hurricane of September 16 to 22, 1938." *Monthly Weather Revue,* vol. 86, pp. 286–88, 1938.

Thomson, B. F. *The Changing Face of New England.* Boston: Houghton Mifflin, 1977.

Webster, N. "On the Supposed Change in the Temperature of Winter." In *A Collection of Papers on Political, Literary, and Moral Subjects,* pp. 119–62. New York: Webster and Clark, 1843.

Zeilinga de Boer, J., and D. T. Sanders. *Volcanoes in Human History: The Far-Reaching Effects of Major Eruptions.* Princeton, N.J.: Princeton University Press, 2000.

Zielinski, G. A., P. A. Mayewski, L. D. Meeker, S. Whitlow, M. S. Twickler. "Potential Atmospheric Impact of the Toba Mega-Eruption ~71,000 Years Ago." *Geophysical Research Letters* 23, no. 8, pp. 837–40, 1996.

MAPS

Dyke, A. S., and V. K. Prest. "Paleogeography of Northern North America 18,000–5,000 Years Ago." Geological Survey of Canada, Map 1703A, 1987.

Radway Stone, J., J. P. Schafer, E. Haley Loudon, M. L. DiGiacomo-Cohen, R. S. Lewis, W. B. Thompson. "Quaternary Geologic Map of Connecticut and Long Island Sound Basin." Scientific Investigations Map 2784, U.S. Department of Interior, Reston: U.S. Geological Survey, 2005.

3. Connecticut's Geologic Treasures: Gems and Ores (pp. 56–82)

PRIMARY SOURCES

Andrews, C. M., and A. C. Bates. *The Charter of Connecticut, 1662*. Tercentary Pamphlet 3. New Haven: Yale University Press, 1934.

Black, R.C. III. *The Younger John Winthrop*. New York: Columbia University Press, 1966.

Booth, J. C. *Report on the Cobalt and Nickel Mines of the Chatham Cobalt Mining Company*. Middletown, Conn.: W. D. Starr and Sons, 1855.

Brainard, John. "Matchit Moodus." *Hartford Mirror*, 1825.

Bylaws of the Silver-Lead Mining Company, Middletown, Conn. New York, Chatterton and Co., 1853.

Chomiak, B. A. "An Integrated Study of the Structure and Mineralization at Great Hill, Cobalt, Connecticut." Master's thesis, University of Connecticut, 1989.

Dillard, A. "Sand and Clouds." *Raritan*, vol. 18, no. 2, pp. 30–40, 1998.

Duffy, D., and G. P. Morago. "Experts Don't Expect Rush for Cobalt Gold." *Hartford Courant*, March 21, 1986.

Fowler, W. S. "The Wilbraham Stone Bowl Quarry." *Massachusetts Archaeological Society Bulletin*, vol. 30, pp. 9–21, 1969.

Francfort, E. "Reports on the Mines of the Chatham Cobalt Mining Company." *News and Advertiser Press* (Middletown, Conn.), 1853.

———. *Report on the Cobalt and Nickel Mines of the Chatham Cobalt Mining Co.* Middletown, Conn.: Starr and Son, 1855.

Hall, F. "Letters from the Valley of the Connecticut River." In *From the East and from the West*. Washington, D.C.: Taylor and Morrison, 1840.

———.*History of Middlesex County*. New York: J. B. Beers and Co., 1884.

Huttson, M. Y. *George Henry Durrie (1820–1863), American Winter Landscapist: Renowned through Currier and Ives*. Santa Barbara, Cal.: Santa Barbara Museum of Art, 1977.

London, D. "Pegmatites of the Middletown District, Connecticut." *New England Intercollegiate Geological Conference Fieldguide*, pp. c5, 1–25, 1985.

Morago, G. P. "Nuggets of Truth behind State's Gold Rush." *Hartford Courant*, April 10, 1986.

Percival, J. G. "Geological Report." In *The Middletown Silver-Lead Mining Company*. New York: Chatterton and Co., 1853.

Shepard, C. U. *A Report on the Geological Survey of Connecticut*. Hartford: State of Connecticut, 1837.

Shepard, O. *Connecticut Past and Present*. New York: Alfred Knopf, 1939.

Stearns, H. T. *Memoirs of a Geologist: From Poverty Peak to Piggery Gulch*. Honolulu: University of Hawaii Press, 1983.

Stiles, E. *Literary Diary* (1787). Ed. F. B. Dexter. 3 volumes. New York: Scribner, 1901.

Wallace, A. "Gold Find in Cobalt." *Middletown Press*, July 12, 1986.

Waters, T. F. *A Sketch of the Life of John Winthrop the Younger*. Cambridge, Mass.: Ipswich Historical Society University Press, 1899.

Whitney, J. D. "Geological Report." In *The Middletown Silver-Lead Mining Company.* New York: Chatterton and Co., 1883.
Wilcox, U. V. "The Lewis Walpole site of Farmington, Connecticut." *Archaeological Society of Connecticut Bulletin,* vol. 35, pp. 5–48, 1967.

SECONDARY SOURCES

Alleman, J. E., and B. T. Mossman. "Asbestos Revisited." *Scientific American,* vol. 277, pp.70–75, July 1997.
Altamura, R. J. "Geology of the Ultramafic Rocks near Westfield, Massachusetts." Master's thesis, Wesleyan University, 1983.
Bartels, O. G. "Radioactive Columbite from Haddam, Connecticut." *Rocks and Minerals,* vol. 28, p. 156, 1953.
Blake, W. P. "Principal Localities of Nickel Ore in America." *U.S. Mineral Resources 1882,* pp. 401–2, 1883.
Boos, M. F., E. E. Maillot, and M. Mosier. "Investigation of Portland Beryl-Mica District in Middlesex County, Connecticut." U.S. Bureau of Mines Reports 4425, 1949.
Bowles, O. "Asbestos." U.S. Bureau of Mines Bulletin 403, 1937.
Brodeur, P. *Outrageous Misconduct: The Asbestos Industry on Trial.* New York: Pantheon, 1985.
Bushnell, D. J. *The Use of Soapstone by the Indians of the Eastern United States.* Washington, D.C.: Annual Report of the Smithsonian Institution, 1939.
Cameron, E. N. "Fluid Inclusions in Beryl and Quartz from Pegmatites of the Middletown District, Connecticut." *American Mineralogist,* vol. 38, pp. 218–62, 1953.
———. "Pegmatite Investigations 1942–45." New England U.S. Geological Survey Professional Paper 225, 1954.
Cameron, E. N., and V. E. Shainin. "The Beryl Resources of Connecticut." *Economic Geology,* vol. 42, pp. 353–67, 1947.
Chomiak, B. A., N. H. Gray, A. R. Philpotts, and R. P. Steinen. "Native Gold at Great Hill, Cobalt, Connecticut." *Geological Society of America,* abstract of the Annual Meeting, 1986.
Dann, K. T. *Traces on the Appalachians.* New Brunswick, N.J.: Rutgers University Press, 1988.
Davis, J. W. "Minerals of Haddam, Connecticut." *Mineral Collector,* vol. 8, pp. 50–54 and 65–70, 1901.
Field, D. D. "A Statistical Account of the County of Middlesex," Middletown, Conn.: 1819.
———. "Middlesex Centennial Address." Middletown, Conn.: Casey, 1853.
———. "Sketch of Chatham." Centennial address. Middletown, Conn.: News and Advertiser Printing Co., 1853.
Fowler, W. S. "Tool-making at the Westfield Steatite Quarry." *American Antiquity,* vol. 11, pp. 95–101, 1945.
———. "Stone Bowl Making at the Westfield Quarry." *Massachusetts Archaeological Society Bulletin,* vol. 30, nos. 1–2, pp. 6–16, 1968.
———. "The Wilbraham Stone Bowl Quarry." *Massachusetts Archaeological Society Bulletin,* vol. 30, nos. 3–4, pp. 9–21, 1969.

———. "The Diagnostic Stone Bowl Industry." *Massachusetts Archaeological Society Bulletin,* vol. 36, nos. 3–4, pp. 1–10, 1975.

Foye, W. G. "Mineral Localities in the Vicinity of Middletown, Connecticut." *American Mineralogist,* vol. 7, pp. 4–12, 1922.

Gass, I. "Ophiolites." *Scientific American,* vol. 247, no. 2, pp. 122–31, 1983.

Gillette, S. G. "Some Minerals of the Gillette Quarry, Haddam Neck." *Rocks and Minerals,* vol. 12, p. 333, 1937.

Graf, A. N. "An Integrated Study of the Great Hill Area, Middlesex County, Connecticut." Master's thesis, Wesleyan University, 1969.

Harte, C. R. "Connecticut's Minor Metals and Her Minerals." Connecticut Society of Civil Engineers Annual Report, no. 61, 1945.

Hayes, A. A. "Notice of Cobalt, Nickel, Etc. of the Chatham Mine, Connecticut." *American Journal of Science,* vol. 21, pp. 195–96, 1832.

Hillebrand, W. F. "On the Occurrence of Nitrogen in Uraninite and on the Composition of Uraninite in General." *American Journal of Science,* vol. 40, pp. 384–94, 1890.

Hinman, R. H. *Connecticut's Part in the American Revolution.* Hartford: E. Gleason, 1842.

Ingerson, E. "Uraninite and Associated Minerals from Haddam Neck, Connecticut." *American Mineralogist,* vol. 23, pp. 269–76, 1938.

Jenks, W. F. "Pegmatites at Collins Hill, Portland, Connecticut." *American Journal of Science,* vol. 30, pp. 177–97, 1935.

Martin, D. S. "Tourmaline Mine at Haddam Neck." *Mineral Collector,* vol. 7, pp. 157–58, 1900.

———. "Beryl from Haddam Neck." *New York Academy of Science Annuals,* vol. 18, pp. 294–95, 1908.

Mather, W. W. "Beryl." *American Journal of Sciences,* vol. 1, p. 242, 1818.

Mohrman, H. W. "Indian Soapstone Quarries and Quarrying Methods in Western Massachusetts." *Illinois State Archaeological Bulletin,* vol. 3, pp. 2–3, 1946.

Nash, A. "Notices of the Lead Mines and Veins of Hampshire County, Massachusetts." *American Journal of Science and Arts,* vol. 14, pp. 238–70, 1827.

Newton, C. C. *Once Upon a Time in Connecticut.* New York: Houghton Mifflin Company, 1916.

Otfinoski, C. "Pegmatite Quarrying in the Middletown District." Unpublished manuscript available at Rockfall Foundation, Middletown, Conn., 1988.

Otis, L. "Stones of Stone Age New England." *Massachusetts Archaeological Society Bulletin,* vol. 11, no. 2, p. 45, 1950.

Pennell, F. W. "On Some Critical Species of the Serpentine Barrens." *Bartonia,* vol. 12, pp. 1–23, 1930.

Rice, W. N. "Minerals from Middletown." *American Journal of Science,* vol. 29, p. 263, 1885.

Rice, W. N., and W. G. Foye. "Guide to the Geology of Middletown and Vicinity." Connecticut Geological and Natural History Survey Bulletin 41, 1927.

Richardson, C. S. "The Cobalt and Nickel Mines in Chatham, Connecticut." *The Mining Magazine,* vol. 2, pp. 124–28. New York: John F. Trow, 1854.

Roberts, G. S. *Historic Towns of the Connecticut River Valley.* Schenectady, N.Y.: Robsen and Adee, 1906.

Russell, H. S. *Indian New England, before the Mayflower.* Hanover, N.H.: University Press of New England, 1980.
Schairer, J. F. "Minerals of Connecticut." Connecticut Geological and Natural History Survey Bulletin 51, 1931.
Scovil, J. A. "The Gillette Quarry, Haddam Neck, Connecticut." *Mineralogical Record,* vol. 23, pp. 19–28, 1992.
Seaman, D. M. "The Walden Gem Mine." *Rocks and Minerals,* vol. 38, pp. 355–62, 1963.
Shannon, E. V. "The Old Cobalt Mine in Chatham." *American Mineralogist,* vol. 6, pp. 88–90, 1921.
Shepard, O. *Connecticut Past and Present.* New York: Alfred Knopf, 1939.
Silliman, B., and C. R. Goodrich. *The World of Science, Art and Industry.* Illustrated examples of cobalt minerals in the New York Exhibition, 1853–54. New York: G. P. Putnam, 1854.
Sohan, J. A. "Connecticut Minerals." Connecticut Geological and Natural History Survey Bulletin 77, 1951.
Stearns, H. T. "The Connecticut Pegmatites in Their Heyday." *Lapidary Journal,* pp. 2266–68, 1980.
Stugard, F. "Pegmatites of the Middletown Area, Connecticut." *U.S. Geological Survey Bulletin* 1042-Q, pp. 613–83, 1958.
Torrey, J. "Chrysoberyl of Haddam." *American Journal of Science,* vol. 4, pp. 52–53, 1822.
Webster, J. W. "Localities of Minerals, Observed Principally in Haddam, Connecticut." *American Journal of Science,* vol. 2, p. 239, 1820.
Whittaker, R. H. "The Ecology of Serpentine Soils." *Ecology,* vol. 35, pp. 258–59, 1954.
Williams, F. H. "Prehistoric Remains of the Tunxis Valley." *Connecticut Quarterly,* vol. 3, pp. 150–66 and 403–23, 1897.
Young, W. R., ed. *An Introduction to the Archaeology and History of the Connecticut Valley Indian.* Vol. 1, no. 1. Springfield, Mass.: Museum of Science, 1969.
Zodac, P. "Minerals of the Strickland Quarry." *Rocks and Minerals,* vol. 12, no. 5, pp. 131–44, 1937.

MAPS

Larrabee, D. M., 1971. "U.S. Geological Survey Map Showing Distribution of Ultramafic and Intrusive Mafic Rocks from New York to Maine." Map 1—676.
"The Chatham Cobalt Mining Co. Map." Easthampton Public Library, 1855.

4. Settlers and Soils in the Central Valley: The Legacy of Glacial Lake Hitchcock (pp. 83–104)

PRIMARY SOURCES

Adams, J. T. In *The Founding of New England.* New York: Atlantic Monthly Press, 1921.
Andrews, C. M. "The River Town of Connecticut." *Studies in Historical and Political Science,* vol. 7, pp. 331–456, Baltimore: Johns Hopkins University Press, 1889.

Antevs, E. "The Recession of the Last Ice Sheets in New England." American Geographical Society Research Series, vol. 11, 1922.

Bell, M. "The Face of Connecticut; People, Geology, and the Land." Connecticut Geological and Natural History Survey Bulletin 110, 1985.

Bradford, W. *Bradford's History of Plymouth Plantation, 1601–1646.* Boston, 1898.

———. *Of Plymouth Plantation, 1620–1647.* New York: Knopf, 1952.

Bragdon, K. *Native People of Southern New England, 1500–1650.* Norman: University of Oklahoma Press, 1996.

Brumbach, J. J. "The Climate of Connecticut." State Geological and Natural History Survey Bulletin 99, 1965.

Cronon, W. *Changes in the Land: Indians, Colonists, and the Ecology of New England.* New York: Hill and Wang, 1983.

Countryman, W. A. "Connecticut's Position in the Manufacturing World." *Connecticut Magazine*, vol. 7, pp. 323–27, 1902.

De Laet, J. "Nieuwe Wereldt, ofte Beschrijvinghe van West-Indien" (1625). In *Narratives of New Netherland, 1609–1664.* Ed. J. F. Jameson, pp. 31–60. New York: Charles Scribner's Sons, 1909.

Duffy, J. "Smallpox and the Indians in the American Colonies." Bulletin of the History of Medicine 25, 1951.

Dwight, T. *Travel in New England.* Ed. B. M. Solomon. Boston: Harvard University Press, 1969.

Eddy, J. A. "The Maunder Minimum: A Reappraisal." *Solar Physics*, vol. 89, pp. 195–207, 1983.

Figuier, L. *La Terre avant le déluge.* Paris: Hachette et Cie., 1864.

Hurd, D. *Profile of the Navigation from Northhampton to Hartford.* New Haven: N. and S. S. Jocelin, 1828 (with map titled "The Farmington & Hampden-Hampshire Canals").

Jackson, C. T. *Report on the Geological and Agricultural Survey of the State of Rhode Island, 1839.* Providence: Cranston and Co., 1840.

Likens, G. E., and M. B. Davis. "Post-Glacial History of Mirror Lake and Its Watershed in New Hampshire." *Verhandlungen der Internationale Vereinigung fur Theoretischeund Angewandte Limnologie*, vol. 19, pp. 982–93, 1975.

Mather, C. *The Short History of New England . . . Boston Lecture, June 7, 1694.* Boston: B. Green, 1694.

Noble-Keegan, K., and W. F. Keegan, eds. *The Archaeology of Connecticut.* Storrs, Conn.: Bibliopola Press, 1999

Pond, J. "Letter to William Pond." In *Letters from New England: The Massachusetts Bay Colony, 1629–1638*, ed. E. Emerson, pp. 64–66. Amherst: University of Massachusetts Press, 1976.

Pratt, J. "Pratt's Apology." *Massachusetts Historical Society Collections* 2, vol. 7, pp. 126–29, 1818.

Ridge, J. C., and F. D. Larsen. "Re-Evaluation of Antevs' New England Varve Chronology and New Radiocarbon Dates of Sediments from Glacial Lake Hitchcock." *Geological Society of America Bulletin* 102, vol. 7, pp. 889–99, 1990.

Roberts, G. S. *Historic Towns of the Connecticut River Valley.* Schenectady, N.Y.: Robson and Adee, 1906.

Shepard, O. *Connecticut Past and Present.* New York: Alfred Knopf, 1939.

Stone, J. R., and G. M. Ashley. "Ice-wedge Casts, Pingo Scars, and the Drainage of Glacial Lake Hitchcock." In P. Robinson and J. B. Brady, eds., *Geology and Geography Conribution*, vol. 2, no. 66, pp. 305–31.

Trumbull, B. *History of Connecticut*, vol. 1. Hartford: Hudson and Goodwin, 1797.

Williams, S. *The Natural and Civil History of Vermont.* Middlebury, Vt.: S. Mills, 1809.

Winthrop, J. "Letter to Sir Simond D'Enes, July 21, 1634." In *Letters from New England*, ed. E. Emerson. Amherst: University of Massachusetts Press, 1976.

Wroth, L.C., ed. *The Voyages of Giovanni de Verrazano, 1524–1528.* New Haven: Yale University Press, pp. 133–43, 1970.

SECONDARY SOURCES

Andersen, R. O. M. *From Yankee to American: Connecticut 1865 to 1914.* Chester, Conn.: Pequot Press, 1975.

Brugam, R. B. "Pollen Indicators of Land-Use Change in Southern Connecticut." *Quaternary Research*, vol. 9, pp. 349–62, 1973.

Buel, R., and J. Bard Mc Nulty, eds. "Connecticut Observed, 1676–1940." *The Acorn Club*, pp. 27–39, 1999.

Byers, D. S. *The Environment of the Northeast.* Andover, Mass.: Phillips Academy, 1946.

Cook, S. F. "The Significance of Disease in the Extinction of the New England Indians." *Human Biology*, vol. 45, pp. 485–508, 1973.

Daniels, B. C. *The Connecticut Town.* Middletown, Conn.: Wesleyan University Press, 1979.

Drake, S. G. *Biography and History of the Indians of North America.* Boston: Antiquarian Institute, 1837.

Field, D. D. *A Statistical Account of the County of Middlesex in Connecticut.* Connecticut Academy of Arts and Sciences, 1819.

Flint, R. F. "The Glacial Geology of Connecticut." Connecticut Geological and Natural History Survey Bulletin 47, 1930.

Gordon, Robert. "Lands Discovered, Lands Transformed in Connecticut: Past, Present, and Future." *Transactions of the Connecticut Academy of Arts and Sciences*, vol. 56, pp. 15–16, 1999.

Hoornbeck, B. "An Investigation into the Cause or Causes of the Epidemic which Decimated the Indian Population of New England, 1616–1619." *New Hampshire Archaeologist*, vol. 19, pp. 35–46, 1967.

Hoyt, D. V., and K. H. Schatten. *The Role of the Sun in Climate Change.* Oxford: Oxford University Press, 1997.

Johnson, F., ed. "Man in Northeastern North America." Papers of the R. S. Peabody Foundation for Archaeology, vol. 3, 1946.

Kavash, B. *Native Harvests.* American Indian Archaeological Institute, Washington, D.C. Depot, Conn.: Shiver Mountain Press, 1977.

Koteff, C. "The Morphologic Sequence Concept and Deglaciation of Southern New England."In *Glacial Geomorphology: Publications in Geomorphology*, ed. D. R. Coates, pp. 121–44. Boston: G. Allen and Unwin, 1982.

Koteff, C., and F. Pessl. "Systematic Ice Retreat in New England." U.S. Geological Survey Professional Paper 1179, 1981.

Lavin, L. "The Morgan Site, Rocky Hill, Connecticut: A Late Woodland Farming Community in the Connecticut River Valley." *Bulletin of the Archaeological Society of Connecticut* 51, pp. 7–21, 1988.

Lewis, T. R. "The Landscape and Environment of the Connecticut River Valley." In *The Great River: Art and Society of the Connecticut Valley, 1635–1820.* Hartford: Wadsworth Atheneum, 1985.

McWeeney, L. J. "Archaeological Settlement Patterns and Vegetation Dynamics in Southern New England in the Late Quaternary." Ph.D. diss., Yale University, 1994.

Nichols, G. E. "The Vegetation of Connecticut II: Virgin Forests." *Torreya* 13, pp. 199–215, 1913.

O'Shea, M. "Springfield's Puritans and Indians." *Historical Journal of Massachusetts*, vol. 26, no. 1, pp. 46–72, 1998.

Perry, C. E. *Founders and Leaders of Connecticut, 1633–1783.* Boston: Heath and Company, 1934.

Roth, D. M. *Connecticut: A Bicentennial History.* New York: Norton and Company, 1979.

Russell, H. S. *Indian New England before the Mayflower.* Hanover, N.H.: University Press of New England, 1980.

———. *A Long, Deep Furrow: Three Centuries of Farming in New England.* Hanover, N.H.: University Press of New England, 1976.

Thomas, P. A. "The Fur Trade, Indian Land and the Need to Define Adequate Environmental Parameters." *Ethnohistory*, vol. 28, pp. 359–79, 1981.

Thomson, B. F. *The Changing Face of New England.* New York: MacMillan, 1985.

Van Dusen, A. E. *Connecticut.* New York: Random House, 1961.

———. "Connecticut, 1763–1818." In *Connecticut History and Culture*, ed. D. M. Roth, 1985.

Wood, W. *New England's Prospect: A True, Lively and Experimental Description.* London, 1634. Reprinted Boston, 1865.

5. The Metacomet Ridge: The Scientific, Political, and Cultural Impact of an Old Lava Flow (pp. 104–131)

PRIMARY SOURCES

Barber, J. W. *Connecticut Historical Collections: History and Antiquities of Every Town in Connecticut.* New Haven: B. L. Hamless, 1836.

Desmarest, N. *Mémoire sur l'origine et la nature du basalte.* Paris, 1774.

Drake, S. G. *Biography and History of the Indians of North America.* Boston: Antiquarian Institute, 1837.

Field, M. E. The Home Book of the Picturesque; or, American Scenery, Art, and Literature. New York: Putnam, 1852.

Fisher, G. P. *The Life of Benjamin Silliman.* 2 vols. New York: Charles Scribner and Co., 1866.

Franklin, B. *The Complete Works of Benjamin Franklin.* Ed. J. Bigelow. New York: G. P. Putnam's Sons, 1888.

Galton, P. M. Prosauropod Dinosaurs of North America. *Postilla,* no. 169, pp. 1–98, 1976.

Hitchcock, E. "A Sketch of the Geology, Mineralogy, and Scenery of the Region Contiguous to the River Connecticut." *American Journal of Science,* vol. 6, pp. 1–86 and 201–36, and vol. 7, pp. 1–30, 1823–24.

———. *Report on the Geology, Mineralogy, Botany, and Zoology of Massachusetts.* Amherst: Commonwealth of Massachusetts, 1833.

Hutson, M. Y. *George Henry Durrie (1820–1863).* Santa Barbara Museum of Art, 1977.

———. *George Henry Durrie (1820–1863): American Winter Landscapist.* Orefield, Pa.: America Art Review Press, 1977.

Hutton, J. *Theory of the Earth.* Royal Society of Edinburgh, 1785.

Kelly, F. *Frederic Edwin Church and the National Landscape.* Washington, D.C.: Smithsonian Institution Press, 1988.

Kistler, T. M. *The Rise of Railroads in the Connecticut River Valley.* Northampton, Mass.: Smith College Studies in History, 1938.

Olsen, P. E. "A 40-Million-Year Lake Record of Early Mesozoic Orbital Climatic Forcing." *Science,* vol. 234, pp. 842–48, 1986.

Olsen, P. E., N.G. McDonald, P. Huber, and B. Cornet. "Stratigraphy and Paleoecology of the Deerfield Rift Basin (Triassic-Jurassic, Newark Supergroup), Massachusetts." *Guidebook,* vol. 2, pp. 488–510. Amherst: University of Massachusetts, 1992.

Paley, W. *Natural Theology.* Hallowell, Maine.: E. Goodale, 1819.

Pynchon, W. H. C. "The Ancient Lavas of Connecticut." *Connecticut Quarterly,* vol. 2, no. 4, pp. 310–19, 1986.

———. "The Black Dog." *Connecticut Quarterly,* vol. 4, pp. 153–61, 1898.

Robinson, W. F. "Abandoned New England." *New York Graphic Society,* pp. 18–36, 1976.

Ruskin, J. *Modern Painters.* New York: E. P. Dutton, 1906.

Shepard, O. *Connecticut: Past and Present.* New York: Alfred Knopf, 1939.

Silliman, B. "Sketch of the Mineralogy of the Town of New Haven, Connecticut." *Connecticut Academy of Arts and Sciences Memoir,* vol. 1, pp. 83–96, 1810.

———. *Remarks Made on a Short Tour between Hartford and Quebec in the Autumn of 1819.* New Haven: Converse, 1820.

———. "Natural Ice Houses." *American Journal of Science,* ser. 1, vol. 3, p. 365, 1822.

———. "Note on the Geological Position of West Rock." *American Journal of Science,* ser. 1, vol. 8, p. 1, 1824.

———. "Igneous Origin of Some Trap Rocks." *American Journal of Science,* ser. 1, vol. 17, pp. 119–32, 1830.

———. *Consistency of the Discoveries of Modern Geology with the Sacred History of the Creation and the Deluge.* New Haven: Hezekiah Howe and Co., 1833.

Twain, M. (1877). In S. C. Wadsworth, "The Towers of Talcott Mountain." *Connecticut Quarterly,* vol. 1, pp. 180–86, 1895.

Werner, A. G. *Von den Ausserlichen Kennzeichen der Fossilien,* 1774.

Whittier, J. G. "Monte-Video." *Connecticut Quarterly,* vol. 1, p. 187, 1895.

SECONDARY SOURCES

Anonymous. *New Hampshire General Court: An Act to Incorporate the Connecticut River Canal Company.* State of New Hampshire, 1828.

Anonymous. *Massachusetts General Court: An Act to Incorporate the Proprietors of the Central Locks and Canals of the Connecticut River.* Boston, 1828.

Bell, M. "The Face of Connecticut: People, Geology and the Land." Connecticut Geological and Natural History Survey Bulletin 110, 1985.

Bickford, C. P., and J. B. McNulty. *John Warner Barber's Views of Connecticut Towns, 1834–1836.* Hartford: The Acorn Club of the Connecticut Historical Society, 1990.

Buckley, W. E. *A New England Pattern: The History of Manchester, Connecticut.* Chester, Conn.: Pequot Press, 1973.

Castle, H. A. *The History of Plainville, Connecticut, 1640–1918.* Chester, Conn.: Pequot Press, 1972.

Clark, G. L. *A History of Connecticut: Its People and Institutions.* New York: G. P. Putnam's Sons, 1914.

Cronon, W. *Changes in the Land; Indians, Colonists, and the Ecology of New England.* New York: Hill and Wang, 1983.

Drake, S. G. *Biography and History of the Indians of North America.* Boston: Antiquarian Institute, 1837.

Harte, C. R. "Connecticut's Canals." Reprinted from the Fifty-Fourth Annual Report of the Connecticut Society of Civil Engineers, Inc., pp. 4–46, 1938.

Hurd, D. "Profile of the Navigation from Northhampton to Hartford." New Haven: N. and S. S. Jocelin, 1828. (With map entitled "The Farmington & Hampden-Hampshire Canals."

McDonald, N. G. "The Connecticut Valley in the Age of Dinosaurs: A Guide to the Geologic Literature, 1681–1995." Connecticut Geological and Natural History Survey, Hartford, Connecticut, 1996.

McHone, J. G. "Broad-Terrane Jurassic Flood Basalts across Northeastern North America." *Geology,* vol. 24, pp. 319–22, 1996.

———. "Connecticut in the Mesozoic World." Connecticut Geologic and Natural History Survey Miscellaneous Paper, 2004.

Olsen, P. E. "Giant Lava Flows, Mass Extinctions, and Mantle Plumes." *Science,* vol. 284, pp. 604–5, 1999.

Palfy, J., J. K. Mortensen, E. S. Carter, P. L. Smith, R. M. Friedman, and H. W. Tipper. "Timing the End-Triassic Mass Extinction: First on Land, then in the Sea." *Geology,* vol. 28, no. 1, pp. 39–42, 2000.

Writer's Program Connecticut. "Boats across New England Hills." State Department of Education, 1941.

MAPS

"Figure of Basalt Flow in Rocky Hill Quarry," *American Journal of Science,* 1830.

Walter, C. E. "Map of the Hampshire and Hampden Canal," 2006.
Walter, C. E., and R. S. Hummel. "Map of the Farmington Canal," 2000.

6. The Moodus Noises: The Science and Lore of Connecticut Earthquakes (pp. 132–156)

PRIMARY SOURCES

Adams, J. "Postglacial Faulting in Eastern Canada: Nature, Origin and Seismic Hazard Implications." *Tectonophysics*, vol. 163, pp. 323–31, 1989.

Anonymous. "The Moodus Noises." In *The Classics*, p. 483, most likely written by John Johnston, 1841.

Anonymous. *Historical Seismicity of New England.* Boston: Weston Geophysical Research, 1977.

Barber, J. W. *Connecticut Historical Collections.* New Haven: B. L. Hamlen, 1836.

Becquerel, A. H. "Ubereinige Eigenschaften der X-Strahlen des Radiums." *Physikalische Zeitschrift*, vol. 6, pp. 666–69, 1905.

Brainard, J. G. C. *Occasional Pieces of Poetry.* New York: Clayton and Van Norden, Printers, for E. Bliss and E. White, 1825.

Brigham, W. T. *Volcanic Manifestations in New England: Being an Enumeration of the Principal Earthquakes from 1638 to 1869.* Memoirs of the Boston Society of Natural History 2. Boston: Society of Natural History, 1874.

Chauncy, C. "Earthquakes." Boston: Edes, Gill, and Draper, 1755.

Clemons, W. H. "The Legends of the Machimoodus." *Connecticut Magazine*, vol.7, pp. 451–58, 1902/3.

Corcoran, T. H. (transl.). *Seneca: Naturales Questiones.* Cambridge, Mass.: Harvard University Press, 1971.

Dana, J. D. "'Volcanic Manifestations in New England, Etc.,' by W. T. Brigham." *American Journal of Science*, vol. 1, pp. 304–5, 1871.

De Forest, J. W. *History of the Indians of Connecticut.* Hartford: W. J. Hamersley, 1851.

Ebel, J. E., V. Vudler, M. Celata. "The 1981 Micro-Earthquake Swarm near Moodus, Connecticut." *Geophysical Research Letters*, vol. 9, no. 4, pp. 397–400, 1982.

Foye, W. G. "The Geology of Eastern Connecticut." Connecticut Geological and Natural History Survey Bulletin 74, 1949.

Franklin, B. "Conjectures Concerning the Formation of the Earth." *Transactions of the American Philosophical Society*, vol. 3, pp. 1–5, 1793.

Gilbert, G. K. "A Theory of the Earthquakes of the Great Basin, with a Practical Application." *American Journal of Science*, vol. 27, pp. 49–54, 1884.

Goethe, J. W. von. *Faust.* Chicago: Great Books of the Western World, 1952.

Harrington, K. P. *Richard Alsop: A Hartford Wit.* Middletown, Conn.: Mattabessett Press, 1969.

Heck, N. H. *Earthquakes.* Princeton, N.J.: Princeton University Press, 1936.

Hitchcock, E. "Remarks on Fault Movements in a Quarry at Portland, Connecticut." *American Academy of Arts and Sciences Proceedings*, vol. 6, p. 105, 1863.

Hosmer, J. K., ed. *Winthrop's Journal.* New York: Barnes and Noble, 1946.

Jackson, M. "Poor Richard's [Franklin's] Theory of the Earth." *Institute for Rock Magnetism Quarterly*, vol. 3, pp. 1–6, 1996.

Johnston, J. "Notice of Some Spontaneous Movements Occasionally Observed in the Sandstone Strata in One of the Quarries at Portland, Connecticut." *American Association for the Advancement of Science Proceedings*, vol. 8, pp. 283–86, 1855.

Linehan, D. "A Revaluation of the Intensity of the East Haddam Connecticut Earthquake of May 16, 1791." Report for Boston Edison Co., 1964.

Lyell, C. *Principles of Geology*. London: John Murray, 1832.

Michell, J. "Conjectures Concerning the Cause, and Observations upon Phaenomena of Earthquakes." *Philosophical Transactions*, vol. 51, pp. 566–634, 1760.

Milne, J. *Seismology*. London: Kegan, Trench, Trubner, 1898.

National Research Council. *Living on an Active Earth*. Washington, D.C.: The National Academic Press, 2003.

Niles, W. H. "On Some Expansions, Movements and Fractures of Rocks Observed at Monson, Massachusetts." *American Association for the Advancement of Science Proceedings* 22, vol. 2, pp. 156–63, 1869.

———. "The Geological Agency of Lateral Pressure Exhibited by Certain Movements of Rocks." *Boston Society of Natural History Proceedings* vol. 18, pp. 272–84, 1877.

Plant, M. "A Journal of the Shocks of Earthquakes Felt near Newbury in New England from 1727–41." *Philosophical Transactions*, vol. 42, pp. 33–42, 1741/42.

Prince, T. *Earthquakes the Works of God: Tokens of His Just Displeasure*. Boston: Henchman, 1727.

Speck, F. G. "Notes on the Mohegan and Niantic Indians." Anthropological Papers of the American Museum of Natural History, vol. 3, New York, 1909.

Sperry, R. T. "Moodus Noises: A Rhyme for the Fourth of July," pp. 89–98, 1884.

Thompson, W. G. *Sea Level, Climate, and Land Level: Paleoenvironmental Records from the Farm River Marsh, Branford, Connecticut*. Master's thesis, Wesleyan University, 1999.

Twain, M. *The Family Mark Twain: Speech on the Weather*. New York: Harper & Brothers, pp. 1107–11, 1935.

White, H. *Early History of New England*. Boston: Sanborn, Carter, Bozin Co., 1841.

Williams, E. "The Duty of a People, under Dark Provisions, or Symptoms of Approaching Evil, to Prepare to Meet Their God." Sermon, November 23, 1755.

Williams, F. A. A. "Observations and Conjectures on the Earthquakes of New England." *Memoirs of the American Academy of Arts and Sciences*, pp. 260–311, 1785.

Winthrop, J. "A Lecture on Earthquakes." Boston: Edes, Gill, and Draper, 1755.

———. "An Account of the Earthquake Felt in New England and the Neighbouring Parts of America, on the 18th of November, 1755." *Philosophical Transactions*, vol. 50, pp. 1–18, 1757/58.

SECONDARY SOURCES

Bakewell, R. *An Introduction to Geology . . .* Ed. B. Silliman. New Haven: Hezekiah Howe and Co., 1833.

Block, J. W., R. C. Clement, L. R. Lew, and J. de Boer. "Recent Thrust Faulting in Southeastern Connecticut." *Geology*, vol. 7, pp. 79–82, 1979.

Brooke, H. V. "Thunder of the Mackimoodus." *Fate*, pp. 70–79, October 1975.

Brunbaugh, D. S. *Earthquakes: Science and Society*. Upper Saddle River, N.J.: Prentice Hall, 1999.

Clark, C. E. "Science, Reason, and an Angry God: The Literature of an Earthquake." *New England Quarterly*, vol. 38, pp. 340–62, 1965.

Davison, E. L. "The Moodus Noises." *East Haddam Historical Society*, pp. 15–20, 1964.

Demos, J. P. *Entertaining Satan: Witchcraft and the Culture of Early New England.* New York: Oxford University Press, 1982.

Dutton, C. E. *Earthquakes in the Light of the New Seismology*. New York: Putnam's Sons, 1904.

Ebel, J. E. "Statistical Aspects of New England Seismicity from 1975 to 1982 and Implications for Past and Future Earthquake Activity." *Seismological Society of America Bulletin* 74, vol. 4, pp. 1311–20, 1984.

———. "The Seventeenth-Century Seismicity of Northeastern North America." *Seismological Research Letters* 67, vol. 3, pp. 51–68, 1996.

———. "The Cape Ann, Massachusetts, Earthquake of 1755: A 250th Anniversary Perspective." *Seismological Research Letters*, vol. 77, no. 1, pp. 74–86, 2006.

Ebel, J. E., K. P. Bonjer, M. C. Onescu. "Paleoseismicity: Seismicity Evidence for Past Large Earthquakes." *Seismological Research Letters* 71, vol. 2, pp. 283–94, 2000.

Ford, J. V. "The Unsolved Mystery of Moodus." *Coronet*, pp. 23–27, 1976.

Guiness, A. C. *Moodus Noises.* MALS thesis, Wesleyan University, 1985.

Hasegawa, H. S., J. Adams, K. Yamazaki. "Upper Crustal Stresses and Vertical Stress Migration in Eastern Canada." *Journal of Geophysical Research*, vol. 90, B5, pp. 3637–48, 1985.

Markham, F. G. "Volcanic and Seismic Disturbances in Southern Connecticut." *Connecticut Magazine*, vol. 9, pp. 68–74, 1905.

McGaffrey, J. P. "New England Seismicity; 1981–1982." *Earthquake Notes* 54, vol. 2, pp. 69–73, 1983.

Niles, H. B. *The Old Chimney Stacks of East Haddam.* New York: Lowe and Company, 1887.

Parker, F. H. "Moodus Noises." *Connecticut Valley Advertiser*, December 17, 1915, p. 4.

Perry, C. *Underground New England.* Boston: Stephen Day Press, 1939.

Perry, E. L. "The Moodus Earthquakes and the Cause of Earthquakes in New England." *National Academy of Sciences*, pp. 401–4, 1941.

Price, C. F. *Yankee Township, East Hampton, Connecticut.* Citizens' Welfare Club, 1941, reprinted 1975.

Shepard, O. *Connecticut: Past and Present.* New York: Alfred A. Knopf, Inc., 1939.

Shute, M. N. "Earthquakes and Early American Imagination: Decline and Renewal in 18th-Century Puritan Culture." Ph.D. diss., University of California, Berkeley, 1977.

Silliman, B. "Notice of an Earthquake at Hartford, Connecticut." *American Journal of Science*, ser. 1, vol. 32, p. 399, 1837.

———. "Earthquake in Connecticut." *American Journal of Science*, vol. 39, pp. 335–42, 1840.

———. "The Connecticut Earthquake." *American Journal of Science*, ser. 2, vol. 26, p. 298, 1858.

Simmons, J. *Our New England Earthquakes.* Boston: Boston Edison Co., 1977.

Skinner, C. M. *Myths and Legends of Our Own Land.* Philadelphia: J. P. Lippincott Company, 1896.

Stark, B. P. "Witchcraft in Connecticut." In *Connecticut History and Culture*, ed. D. M. Roth, p. 95, 1985.

Sterry, I. H., and W. H. Garrigus. *They Found a Way, 13 (Witches).* Brattleboro, Vt.: Stephen Daye Press, 1938.

Street, R., and A. Lacroix. "An Empirical Study of New England Seismicity; 1737–1977." *Seismological Society of America Bulletin*, vol. 69, no. 1, pp. 159–75, 1979.

Todd, C. B. *In Olde Connecticut.* New York: Grafton Press, 1906.

Trumbull, B. *A Complete History of Connecticut . . .* New Haven: Maltby, Goldsmith and Co., 1818.

Woodworth, J. B. "Post-Glacial Faults of Eastern New York." *New York State Museum Bulletin*, vol. 107, pp. 5–28, 1904.

MAPS

Frankel, A., M. D. Peterson, C. S. Mueller, K. M. Haller, R. L. Wheeler, E. V. Leyendecker, R. L. Wesson, S. C. Harmsen, C. H. Cramer, D. M. Perkins, and K. S. Rukstales. "National Seismic Hazard Map." Open File Report 96–532. Denver: U.S. Geological Survey, 1996.

7. Visitors from Space: The Weston and Wethersfield Meteorites (pp. 157–170)

PRIMARY SOURCES

Bowditch, N. "An Estimate of the Height, Direction, Velocity, and Magnitude of the Meteor, That Exploded over Weston, in Connecticut, December the 14th, 1807." *Journal of Natural Philosophy, Chemistry and the Arts*, vol. 28, pp. 89–219, 1811.

Condon, G. "The Sky Is Falling! (Not Really.) But Watch Out for Space Debris." *Hartford Courant*, March 21, 1997.

Dombrowsky, P. "Target: Wethersfield." *Astronomy Forum*, vol. 5, pp. 21–22,1982.

Di Cicco, D. "Wethersfield Meteorite: The Odds Were Astronomical." *Sky and Telescope*, vol. 65, pp. 118–19, 1982.

Frick, U., E. H. Hebeda, L. Schultz, and P. Signer. "Rare Gases in the Weston Meteorite." *Meteoritics*, vol. 6, no. 4, p. 271, 1971.

Giannitti, K. "A Meteoritic Stone." *The Chronicle Quarterly*, vol. 23, no. 2, pp. 1–3, 2002.

Haar, D. "Meteorite Ends Journey from Space at Yale." *Hartford Courant*, Nov. 9, 1985.

Hand, B. "Exhibit Unites Meteorites." *Hartford Courant*, July 22, 1983.

Mason, B., and H. B. Wiik. "The Composition of the Forest City, Tennasilm, Weston, and Geidam Meteorites." *American Museum Novitiates*, American Museum of Natural History, no. 2220, pp. 1–20, 1965.

Narendra, B. L. "The Peabody Museum Meteorite Collection: A Historic Account." *Discovery*, vol. 13, no. 1, pp. 10–23, 1978.

———. "The Wethersfield Meteorite." *Discovery*, vol. 17, no. 1, pp. 27–28, 1983–84.

Poag, C. W., D. S. Powars, L. J. Poppe, and R. B. Mixon. "Meteoroid Mayhem in Ole Virginny: Source of the North American Tektite Strewn Field." *Geology*, vol. 22, pp. 691–94, 1994.

Poag, C. W., C. Koeberl, and W. U. Reiwold. *The Chesapeake Bay Crater.* New York: Springer-Verlag, 2004.

Preston, D. J. "Wethersfield's Meteorite." *Natural History Magazine*,1984.

Rossignol, J. "2nd Meteorite in 11 Years Falls Here." *Wethersfield Post*, November 2, 1982.

Silliman, B., and J. L. Kingsley. "Account of a Remarkable Fall of Meteoric Stones in Connecticut." Republished from a public paper in the *Connecticut Herald*, pp. 39–57, 1807.

———. "Memoir on the Origin and Composition of the Meteoric Stones Which Fell from the Atmosphere in the County of Fairfield, and State of Connecticut, on the 14th of December, 1807." *Transactions of the American Philosophical Society*, vol. 6, pp. 323–45, 1809.

———. "An Account of the Meteor Which Burst over Weston in Connecticut in December, 1807." *Memoirs of the Connecticut Academy of Arts and Sciences*, vol. 1, pp. 141–51, 1810.

———. "Chemical Examination of the Stones Which Fell at Weston, Connecticut, December 14, 1807." *Memoirs of the Connecticut Academy of Arts and Sciences*, vol. 1, pp. 151–61, 1810.

SECONDARY SOURCES

Bevan, A., and J. de Laeter. *Meteorites.* Washington: Smithsonian Institution Press, 2002.

Collins, G. S., and K. Wunnemann. "How Big Was the Chesapeake Bay Impact? Insight from Numerical Modeling." *Geology*, vol. 33, no. 12, pp. 925–28, 2005.

Grady, M. M. *Catalogue of Meteorites.* Cambridge: Cambridge University Press, pp. 475, 523, 524, 2000.

Harris, R. S., M. F. Roden, P. A. Schroeder, S. M. Holland, M. S. Duncan, E. Albin. "Upper Eocene Impact Horizon in East-Central Georgia." *Geology*, vol. 32, no. 8, pp. 717–20, 2004.

Hoffleit, D. "A Sampling of Meteorites at Half-Century Anniversaries of the Academy." In *Connecticut: Past, Present and Future*, pp. 107–24. New Haven: Yale University Press, 1999.

Koeberl, C., C. W. Poag, W. U. Reiwold, D. Brandt. "Impact Origin of Chesapeake Bay Structure and the Source of the North American Tektites." *Science*, vol. 271, pp. 1263–66, 1996.

Nininger, H. H. *Arizona's Meteorite Crater.* Sedona: American Meteorite Museum, 1956.

———. *Out of the Sky.* New York: Dover, 1959.

Norton, O. R. *Rocks from Space.* Missoula, Montana: Mountain Press, 1998.

Pearl, R. M. "New England Meteorites." *Earth Science*, vol. 29, no. 6, pp. 289–90, 1976.

Poag, C. W, D. S. Powars, L. J. Poppe, R. B. Mixon, L. E. Edwards, D. W. Folger, S. Bruce. "Deep Sea Drilling Project Site 612 Bolide Event: New Evidence of a Late Eocene Impact-Wave Deposit and a Possible Impact Site, U.S. East Coast." *Geology*, vol. 20, pp. 771–74, 1992.

Poag, C. W, E. Mankinen, R. D. Norris. "Later Eocene Impacts." In *From Greenhouse to Icehouse*, ed. D. R. Prothero et al., pp. 495–510. New York: Columbia University Press, 2003.

Sanford, W. E., G. S. Gohn, D. S. Powars, J. W. Horton, L. E. Edwards, J. M. Self-Trail, R. H. Morin. "Drilling the Central Crater of the Chesapeake Bay Impact Structure: A First Look." *EOS Transactions of the American Geophysical Union*, vol. 85, no. 39, pp. 369 and 377, 2004.

Schultz, L., P. Signer, J. C. Lorin, and P. Pellas. "Complex Irradiation History of the Weston Chondrite." *Earth and Planetary Science Letters*, vol. 15, pp. 403–10, 1972.

Weston Historical Society. "A Meteoric Stone." *The Chronicle Quarterly*, vol. 23, no. 2, pp. 2–3, 2002.

Index

Page numbers in *italics* refer to figures.

Acadian orogeny, 15
achondrites, 167
Adams, John, 83, 102
African tectonic plate, 15, 21–22
age of Earth, calculating, 10–14, 168
Agnes, Hurricane, 33
agriculture: Central Valley, 3, 83, 89, 90–91, 95–96; cleared land's effects, 2, 34, 49–50; climatic effects on, 35, 41–42, 43–44; colonial, 2–3, 99–100; Connecticut's ascendancy in, 3, 100; glacial till problem for, 49–50; and horticulture by native people, 92–93, 96–98
Albedo, 40
Alleghenian orogeny, 15, 21
Allen, Ethan, 69
Alsop, Richard, 133–34, 144–45
American Gem Company, 73
American Indians. *See* Native Americans
American Revolution, 3, 65–70, 101–2
Anchisaurus dinosaurs, 110–11, 117–18
Andre, Carl, 172n1
Anighito meteorite, Greenland, 168
Antevs, Ernst, 88
Appalachian range, 14, 20, 21–22
aquamarine, 72
Archaic age, 76–78
Aristotle, 145
art and geology, 9, 16, 124–28, 130
asbestos, 81–82
asteroids, threat to Earth from, 170
astroblemes, 166, 168–70
asymmetric flow fold, *18*
Atlantic City, New Jersey, meteorite crater off coast of, 168
Atlantic Ocean, tectonics of, 21, 84, 117
aurora borealis and sunspot activity, 44
Avalon terrane, 26

Bakewell, Robert, 143
Barber, John Warner, 115–16, 126, *127*, 128, 130–31, 143
barite mining, 57
Barndoor Hills, 111
Barringer (Meteor) Crater, Arizona, 166
Bartlett, Mr., 131
basaltic magmas, 6, 15, 23, 84, 125–26. *See also* Metacomet Ridge
Bashful Lady Cave, 172
Becquerel, Henri, 12, 137
Bell, Michael, 28, 45, 83–84
Berkshire Hills, 20
beryls, 70–75
Black, Robert, 57–58
Black Dog, curse of, 129
black stem rust, 100
Blake, Eli Whitney, 120
Block, Adrian, 43, 93, 126

Blodget, Lorin, 34–37
Boltwood, Bertram, 12
Booth, James Curtis, 64
Bowditch, Nathaniel, 164
Bradford, William, 28, 94
Bragdon, Kathleen, 92
Brainard, John, 153–54
Brigham, William, 134, 145
Brown, Edmund, 63
brownstone (sandstone): artistic depiction of, 116, 126; in Central Valley, 85, 115, *116*; quarries for, 3, 57, 82, 172
Buffon, Comte de (Georges-Louis Leclerc), 10–11, 49
Bulkly, Peter, 134
Burr, Merwin, 161–62
Byron, Lord, 36–37

Cambridge, Connecticut, 95
Cameron fault zone, 26
canals, 6, 120–24
Candee, George Edward, 130
Cape Ann earthquake, 137–38
carbonaceous chondrites, 167
carbuncles and earthquake theories, 136–37
Carnian age, 53
Carol, Hurricane, 33
Cassarino, Paul, 157
Castle Craig, 131
Cat Hole Pass, 126, *127*
cattle operations, colonial, 94, 102
Cave Hill, 133
Cenozoic period, 12, *24*, 170. *See also* ice ages
Central Valley: agriculture in, 3, 83, 89, 90–91, 95–96; climate in, 90–91; European settlement, 93–100, *101*; glacial Lake Hitchcock, 85–91; industrialization effects, 103–4; lava flows as basis of, 6, 15, 23; Mesozoic, *24*, 84–85; native settlement, 92–93, 96–98; overview, 83–84; Revolutionary period, 101–2; as rift zone, 22, *24*; sandstone in, 85, 115, *116*; sedimentary deposits, 23, 51–52, 84, 87–88; stratigraphic simplicity of, 17
Chapman, Henry, 136
Char, Lake, 26
Charles II, King of England, 57, 59
Chatham area cobalt mining, 62–64
Chatham Cobalt Mining Company, 64
chert, 76, *77*
Chesapeake Bay meteorite crater, 168–70
chinaware, cobalt for, 62, 64
chondrites, 166–67
chondrules, 166–67
Church, Frederic, 128
clays in glacial Lake Hitchcock, 87–88
cleared land, effects of, 1–2, 34, 49–50
Clemmons, William, 135
climate and weather: ancient climates, 51–55; Archaic age global warming, 76; Central Valley vs. Highlands, 90–91; cleared land's effect on, 2, 34; climate cycles, 38–51; drought conditions, 2, 35, 40, 76; economic effects of, 36, 41–42; historical shifts in, 33–38; hurricanes, 6, 28–33; lava flow effects on, 112; meteorite impact effects on, 170; overview, 27–28; tornadoes, 33
cobalt, 61–64
Cobalt, Connecticut, 56
Cochegan Rock, 47
Cole, Thomas, 125, 128
collecting, rock and mineral, 70–75, 173
collision, tectonic plate, *18*, *25*, 26
Colonial period: agriculture in, 2–3, 99–100; cattle operations, 94, 102; Central Valley habitation, 93–100, *101*; climate adjustment, 33–34, 43; earthquakes during, 134–41; gold find, 57–61; hurricanes, 28; settlement growth in New

England, *97*; water transportation competition, 120–23, *122–23*
columbite (niobium), 75
comets, threat to Earth from, 170
compression, tectonic plate, *18*, *25*, *26*
Cone, Joe, 5
Connecticut charter, expansion of, 58–59
Connecticut River: and agriculture in Central Valley, 3; colonial settlement along, 92, 95–96, 97–98, 99–100; ecosystem types along, 92; post-glacial development, 90; as recreational highway, 5; and water transportation competition, 120–24, *122–23*
Connecticut River Company, 122
Cooper, Thomas, 115
copper mining, 3, 57
Countryman, William, 103–4
Cowper, William, 9
craters, meteorite, 166, 168–70
crustal depression from ice ages, 50, 89–90
crustal extension faulting, *18*
Cryptozoic period, 12
cultural effects of geology, 8–9, 78–80, 120–23, *122–23*. *See also* religious belief
curse of the Black Dog, 129

Dana, James Dwight, 145
"Darkness" (Byron), 36–37
Deane, Silas, 66, 67, 68
De Collibus, Robert, 160
De Forest, John William, 132
deforestation, 1–2, 34, 49–50
Desmarest, Nicolas, 113–14
Devonian period, 20–21
Diane, Hurricane, 33
Dillard, Annie, 61
dinosaurs, 110–11, 117–18
Dinosaur State Park, 172
discordance, geologic, 18–19
Dobson, Peter, 47, 49
Dodd's Granite Quarry, *4*
Donahue, Mr. and Mrs. Robert, 157, *158*
Donna, Hurricane, 33
drought conditions, 2, 35, 40, 76
drumlins, 47
Durant, Will, 9
Durrie, George Henry, 64, 128, 130
Dutch settlers, 93, 96
Dwight, Timothy, 11, 30

earthquakes: early hypotheses, 137–41; epicenters of, *138*, *142*; and Moodus poetry, 153–55; overview, 6, 132–34; pseudo-scientific ideas, 136–37; religious response to, 134–41; responsible forces, 148–53; scientific examination of causes, 141–47
Eastern Highlands, 26
Easton, meteor strike near (1807), 161–65
East Rock, 125–26, 128, 130
ecology, 2, 108–9, 172
economic conditions: climate and weather effects, 36, 41–42; industrialization, 3–5, 81–82, 103–4; trading community in nineteenth century, 101–2; water supply and lava flows, 118–20; water transportation competition, 120–23, *122–23*. *See also* agriculture; mining; quarries
elbaite (tourmaline), 73–74
electric discharge hypothesis for earthquakes, 144
emeralds, 72, 74
Enfield Canal, *122*
Eocene/Oligocene period boundary, meteorite strike at, 170
Erie Canal, 121
Erkelens, Gominus, 61, 62–63
erosion: and colonial agricultural practices, 100; glacial till, 22–23; of lava flows, 108; and suburban/urban development, 4–5, 22–23, 171–72

erratics, glacial, 47, *48*, 49
Estuarine zone, Connecticut River, 92
European vs. New England climate, 33–34, 36
exhaustion of soil, 100
extinctions, 53–54, 113

Face of Connecticut, The (Bell), 28, 83–84
Fahrenheit, Gabriel, 28n1
farming. *See* agriculture
Farmington Canal, 6, *122*
fault blocks near Meriden, 106
faulting: crustal extension, *18*, 23, *25*, 26; and earthquake activity, 146–47, 151; Meriden, 106; overview of Connecticut, 20–21; and rift valley creation, 21–22
Faust (Goethe), 146
feldspar, 3, 64, 73
Field, Mary, 128
folding by compression, *18*, *25*, 26
fool's gold, 163
forest, loss of old-growth, 1–2, 34, 49–50
Fowler, William, 78, 79
Foye, Wilbur, 148–49
Francfort, Eugene, 63–64
Franklin, Benjamin, 34, 145

Gaillard, Lake, 118
galena (lead/silver deposit), 65–70
Gardner, Mrs., 164
gemstones, 70–75
geology: mapping, 14–17, *18*; outcrops overview, 18–23; protection of features, 172; time scale development, 8–14. *See also* earthquakes; glacial activity; tectonics, plate; volcanic activity
Gibbs, Col. George, 163
Gilbert, Karl, 146
Gillette Quarry, 73–74
glacial activity: crustal depression from, 50, 89–90; and earthquake theories, 148–49; effect on landscape, 75–76; and erosion of lava flows, 108; history of Connecticut, 44–51, 55; and Lake Hitchcock, 50, 85–91; sedimentary deposits, 23, *24*, 85–86; and stone walls, 49, 171. *See also* ice ages
glacial erratics, 47, *48*, 49
glacial soils, 85–91
glacial till, 22–23, 49–50
Glastonbury pegmatites, 13
Gloria, Hurricane, 28, 33
Glory and the Dream, The (Manchester), 32
Goede Hoop, Huys de, 96
Goethe, Johann Wolfgang von, 146
gold, 57–61, 64–65, 163
Gondwana continent, 20
granite resource, 3, 56, 57
Great Hill, 60–64
"Great Unconformity," 19, 22, 172
Green Mountains, 20
Gregory, Herbert, 16
gunpowder for Revolution, 68

Haddam, Connecticut, 71–72, 74, 135
Hale, Matthew, 172
Hall, Frederick, 70–72, 81
Hamlin, Jabez, 67
Hampden basalt, *107*, 111, 115, 126
Hampsted, Joshua, 43
Hanging Hills, 126, 129, 131
Hard, Walter, 96
Harte, Charles, 56
Hartford, Connecticut, 6, 95, 96, 120–24
Heck, N. H., 150
Hedden, Stan, 160
heliodor, 72, 73
hematite, 52
Highlands, 17, 26, 47, 49, 90–91
Hillhouse, James, 121
Hitchcock, Edward, 86, 114, 115, 147
Hitchcock, glacial Lake, 50, 85–91
Hobart, Jeremiah, 135
Hobbamocko, 132, 135
Holmes, Oliver Wendell, 31
Holmes, William, 95

Holyoke lava flow: artistic depictions of, 126, *127*; original latitude of, 116–18; structure of, 106, *107*, 108; timing of, 111
Homestead, Iowa, meteor strike, 160
Honey Hill fault zone, 26
Hooker, Thomas, 95–96
horticultural system, native people, 92–93, 96–98
Hosmer, Stephen, 136, 137
Hosmer, Titus, 67
Howe, Henry, 126
Hubbard, William, 131
Hudson, Henry, 43
Hudson Highlands, 20
Hudson River School, 125
Hueblein, Gilbert, 131
humanity: geological influence on, 8–9, 78–80, 120–23, *122–23*, 171; influence on landscape, 171–72; Toba volcano and bottleneck in, 45. *See also* economic conditions
Hunt, Seth, 63
Huntington, meteor strike near (1807), 161–65
hurricanes, 6, 28–33
Hutton, James, 10, 18–19, 113
hydro power, 3

Iapetus Ocean, 20, 25–26
ice ages: and Central Valley, 23, *24*, 85–86; Little Ice Age, 40–44; and orbital eccentricity cycles, 38–39; story of, 44–51, 55; temperatures during, *41*, 44–45. *See also* glacial activity
"icehouses" in basalt, 118–19
igneous (volcanic) rocks: definition, 8; granite, 3, 56, 57; as lithic tool material, 76–77; pegmatite, 12, 13, 20–21, 70–75. *See also* lava flows
Indians, American. *See* Native Americans
industrialization, 3–5, 81–82, 103–4
inventors and entrepreneurs, 4, 103, 104
iron meteorites, 167, 168
iron mining, 3, 57, 59–60

jasper, 76
Jefferson, Thomas, 164
Jocelyn, Nathaniel, 130
Johnson, Emory, 144
Johnson, John, 142
Johnston, John, 147
Judges Cave, *48*
Jurassic period, 22, *53–54*, 113

Kellogg, Mr., 131
Kelvin, Lord, 10–11
King, Clarence, 14–15, 16
King Philip (Metacomet/Pometacom), 105, *106*
King Philip's War, 100
Kingsley, James, 161–64
Knapp, Michelle, 159
Knool, John, 62
Knox, Henry, 66
Kunz, George, 73

Laet, Johannes de, 92
Lake Char, 26
Lake Gaillard, 118
Lake Hitchcock, glacial, 50, 85–91
Laki fissure in Iceland, 112
landscape: conservation efforts, 172; human transformation of, 171–72; overview, 1–7; post-ice age, 75–76; variety of rock types in, 9. *See also* geology
land-use patterns. *See* agriculture; suburban/urban development
Laurasian continent, 20
lava flows, 6, 15, 23, 84, 125–26. *See also* Metacomet Ridge
lead mining, 65–70
Lightfoot, John, 10
Linehan, Daniel, 150–51
Little Ice Age, 40–44
Long Island, New York, 32
Lowell, John, 138

machine-tool industry, 103
magnetic field of the Earth, 40, 116–18
Manchester, William, 32
Manhan lead vein, 69

Manicouagan crater, Quebec, 53
Mansfield, Calvin, 35
Marshall, Herbert, 129
Mason, Capt. John, 98
mass extinctions, 53–54, 113
"Matchit Moodus" (Brainard), 153–54
Mather, Cotton, 34, 120
Mather, Increase, 29
Mattabassett Silver-Lead Mining Company, 69
Maunder Minimum, 44
Medieval Warm Period, 42
Meriden, Connecticut, 106, 126, *127*, 129, 131
Mesozoic period: catastrophic starting point for, 52–53; climate of, 51–52; Connecticut structures from, *24–25*, 26, 84–85; definition, 12; Jurassic period, 22, 53–54, 113; tectonic developments, 21–22; Triassic period, 53–54, 113
Metacomet (King Philip/Pometacom), 105, *106*
Metacomet Ridge: artistic depiction of, 124–28, 130; geologic setting, 109–13; as holder of water supply, 118–20; overview, 105–9; and Plutonist vs. Neptunist theories, 113–16; timing and latitude of formation, 116–18; tower-building on, 130–31; and water transportation systems, 120–23, *122–23*
metal resources, 15, 57–70, 163
metamorphic rocks, 8, 10, 17n4, 20
Meteor (Barringer) Crater, Arizona, 166
meteors and meteorites: and age of Earth, 13, 168; chemical types, 167; definitions, 157n1; impact effects, 53, 157–65, 168–70; incandescence origin, 159n2; monetary value of, 163n4; origins of, 166–68; overview, 6, 165–66; speed of, 161
meteor showers, 166
mica, sheet, 73
Mid-Atlantic Ridge, 150
Middlesex, Connecticut, 56, 70–75
Middletown, Connecticut, 65–70
migration west from New England, 35–36
Milankovitch, Milutin, 38
milky-white quartz, 77–78
Mills, 3, 103
Milne, John, 146
Mine Brook, 61
mineral resources: barite, 57; beryls, 70–75; at Great Hill, *62*; hematite, 52; metals, 15, 57–70, 163; quartz, 20, 77–78; serpentine, 80–82; sheet mica, 73; sulfur, 68
mining: barite, 57; beryls, 70–75; economics of, 56–57, 59–63, 68, 69; major operations, *5*, 82; metals, 3, 57–70; overview, 3, 57; soapstone, 77–80, 82; ultramafics, 80–82; vein quartz, 77–78
Mitchell, John, 145
Modern Painters (Ruskin), 125
"Monte Video" (Whittier), 108
Moodus noises: at Cave Hill, 133; Hosmer's description of, 137; poetry on, 153–55; religious perspective, 132–33, 136; search for natural causes, 143–44. *See also* earthquakes
"Moodus Noises, a rhyme for the fourth of July" (Sperry), 154–55
Morgan, J. P., 73
Morgan archaeological site, 97–98
Morganite, 73
Morse, Jedidiah, 2
Morse, Samuel, 126
mountain-building periods (orogenies), 15, 21
Mount Tom, *133*
Murchison, Sir Roderick, 49

Narendra, Barbara, 157, 159
Native Americans: adaptation to climate changes, 43; Central Valley habitation, 92–93, 96–98;

geological advantages for, *22*; horticultural system of, *92–93*, *96–98*; Lake Char name, *26*; lithic resources for, 76–80; Metacomet/King Philip, 105, *106*; and Moodus noises, 132–33, 135
Nature Conservancy, 172
Neptunists vs. Plutonists on geologic processes, 18–19, 113–16
Newark basin, 112
"New England Express" (1938 hurricane), 31–33
New Haven, Connecticut, 6, 120–24
New Haven-Northampton canal, 122–24
Newtown (Cambridge), Connecticut, 95
nickel mining, 64
Niles, William, 147
1938 hurricane in New England, 31–33
niobium (columbite), 75
North American tectonic plate, 14, 20, 21–22, 117, 150–53
Northampton, Massachusetts, 121–24, *123*
Nowell, Thomas, 100, *101*
Nutmeggers, 56–57

"Old Hole," The, in Dodd's Granite Quarry, *4*
Old Newgate Copper mine, 82
Oligocene/Eocene period boundary, meteorite strike at, 170
Olmstead, Denison, 31, 166
Olsen, Paul, 52
orbital eccentricity and climate change, 38, 55
Ordovician period, 19–20
orogenies (mountain-building periods), 15, 21

Pace, Ted, 160
Page, William, 164
Paleozoic period: catastrophic endpoint for, 52–53; climate information for, 54; Connecticut structures from, 19–21, *24–25*, *26*; definition, 12; mountain-building in, 15
Palisades sill, 112
Pangaea supercontinent, 21, 109
Panthalassa ocean, 21
Parsons, Elijah, 142
Parsons, Samuel Holden, 66
partial melting, 20
Peary, Robert, 168
Peekskill, New York meteor strike, 159, *160*
pegmatite, 12, 13, 20–21, 70–75
Percival, James, 15–16, *17*, 69
Permian period, 21
"pet rock" phenomenon, 173
Pittsburgh, Pennsylvania meteor shower, 166
placer deposit of gold, 61
Plant, Matthias, 137–38
plant development, post-glacial, 88–89
plate tectonics. *See* tectonics, plate
Plattes, Gabriel, 65
Playfair, John, 19
Pleistocene glacial deposits, *24*. *See also* ice ages
Plutonists vs. Neptunists on geologic processes, 18–19, 113–16
plutons, granite, 21
Pocumtucks, 98
poetry: on Connecticut River, 5; on geologic time, 9; on Moodus noises, 132, 144–45, 153–55; on Talcott Mountain, 108; on weather, 31, 36–37
political conditions, 36, 96
Pometacom (Metacomet/King Philip), 105, *106*
Popigai Crater, Siberia, 169
population increase in eighteenth century, 35–36
porcelain, cobalt for, 62, 64
Poronall, Thomas, 102
Porter, Ephraim, 163
Portland brownstone quarries, 3, 57, 82, 172

Preston, Douglas, 159
Price, C. F., 144
Primitive Origination of Mankind, The (Hale), 172
Prince, Thomas, 144
Prince, William, 162
Puritans. *See* Colonial period
Pynchon, William, 95–96, 98, 105, 129

quarries: basalt, 120; brownstone, 3, 57, 82, 172; for farm foundation stones, 171; gems in tailings of, 73–74; granite, *4*, 57; historical overview, 3, 5; major operations, *5*, 82; preservation of historic, 172; remains of, *56*; soapstone, 78–80
quartz, 20, 77–78

radioactive decay, 12, 137
recession, glacial, 47
Redfield, William C., 31
red sedimentary beds in Central Valley, *51–52*
Reed, Abner, 126
religious belief: conflicts among settlers and migration, 95, 96; and earthquakes, 134–41; and geologic process explanations, 113, 114–15; and meteor strikes, 165; and Moodus noises, 132–33, 136; New Haven/Hartford conflict, 120; soapstone bowls as sacred objects, 79; and suppression of art by Puritans, 125; and time scale of the Earth, 10–11
reservoirs, 118
Revolutionary period, 3, 65–70, 101–2
Rhaetian age, *53*
rifting process, 21–22, *24*, *25*, 84
riverine zone, Connecticut River, 92, 95–96, 97–98, 99–100
Roaring Brook, 19–20, 22–23
Roberts, George, 83
Robinson, Henry, 16
rocks, importance for humans, 8–9, 78–80, 120–24, *122–23*, 171. *See also* geology
Rocky Hill, 86, 88, 115, 117–18
Rodgers, John, 16
Rose, Heinrich, 75
Ruskin, John, 125
Russell, Howard, 100
Russell, Capt. Samuel, 67
Rutherford, Ernest, 12

Salisbury Iron District, 57, 82
saltpeter, 68
Samarskite, 75
Sand and Clouds (Dillard), 61
sandstone. *See* brownstone
schists, Roaring Brook, 20
sea level changes and ice ages, 50
sedimentation: in Central Valley, 23, 51–52, 84, 87–88; glacial and post-glacial, 23, *24*, 85–86; process of, 10
Seely, Elijah, 163
seismic activity. *See* earthquakes
Seneca, 139–40
"September Gale, The" (Holmes), 31
serpentine, 80–82
1791 earthquake, 141–42, *143*
sheet mica, 73
Shepard, Charles, 15–16
Shepard, Odell: on artistic depiction of Connecticut, 124; on Connecticut inventors, 104; on expansion of Connecticut's charter, 59; on geology's effect on human activity, 8; on layout of colonial Connecticut, 102; on stone walls, 49; synopsis of Connecticut, 1
Siccar Cliff, Edinburgh, Scotland, 18–19
Silliman, Benjamin: artistic description of geology, 125; geologic mapping, 15; and geologic time calculation, 11; on meteor dangers, 164–65; meteor

strike documentation by, 161–64; Plutonism of, 114, 115
Silurian period, 19–20
silver mining, 66–69
Skin of Our Teeth, The (Wilder), 51
Sleeping Giant, 111, 130
smallpox epidemic among native people, 94
smalt, 62
soapstone (steatite), 77–80, 81, 82
soils, 2, 4, 8, 85–91, 100
solar energy fluctuations and climate change, 38, 44
Sperry, Reginald, 132, 154–*55*
spin axis wobble and climate change, 38
Springfield, Connecticut, 95
Staples, Elihu, 163
steatite (soapstone), 77–80, 81, 82
Steele, Dr., 136
Stephauney, John Sebastian, 61–62
Stiles, Ezra, 28, 50, 60
Stone by Stone (Thorson), 49
stone walls and glacial deposits, 49, 171
Stony Creek, 56
Stony Creek Granite District of Branford and Guilford, 82
Stratford meteor strike (1974), 167
stratigraphy, 16–17
Strickland Quarry, 73, 74, 172
subduction, 61
Subterranean Treasures (Plattes), 65
suburban/urban development, 2, 4–*5*, *22–23*, 119–20, 171–72
Sudbury astrobleme, Ontario, Canada, 166
sulfur, 68
sunspot activity and climate change, 44
Sylacauga, Alabama meteor strike, 159

Taconic Mountains, 19–20
Talcott, Matthew, 67
Talcott lava flow, *107*, 111
Talcott Mountain, 105, 108, 131
talus (basalt blocks), 108
Tambora volcano, 37
Tashua (Tashowa) Hill, meteor strike at, 163
tectonics, plate: and climate change, 40, *55*; collision, *18*, *25*, 26; and Connecticut's development, 14, 20, 21–26; framework for, 23–26; and Metacomet Ridge formation, 109, 116–18; and Moodus-area earthquakes, 150–*53*; New England's history in, 19–20; post-glacial, 149–50; rifting process, 21–*22*, *24*, *25*, 84; subduction, 61; and volcanism, 15. *See also* faulting
tektites from meteorite impact, 170
temperatures: Highlands vs. Central Valley, *91*; ice age, *41*, 44–*45*; New Haven means, 28, *29*; nineteenth century, 34–38
terminal moraine, 47
textile mills, 103
Theophilus of Antioch, 10
Theory of the Earth (Hutton), 10
thermohaline circulation in oceans, 39
Thompson, William, 149–50
Thoreau, Henry David, 125
Thorson, Robert, 49
Tiffany's, 73
till, 22–23, 47, 49–50
tilt, gorge of, 19
time scale, geologic, 8–14
Toba volcano, 45
Tom, Mount, *133*
tornadoes, 33
Totoket Mountain, 120
tourmaline (elbaite), 73–74
Toutatis asteroid, 170
tower-building on Metacomet Ridge, 130–31
trading community, nineteenth-century, 101–2
transportation systems, 6, 120–23, *122–23*
traprock (basalt), 6, 15, 23, 84, 125–26. *See also* Metacomet Ridge

Triassic period, *53–54*, 113
Trumbull, Jonathan, 95, 101
Tunxis Valley, 79
Twain, Mark, 27, 131, 153

ultramafics, 80–82
uplands zone, Connecticut River, 92
urban/suburban development, 2, 4–5, 22–23, 119–20, 171–72
Ussher, James, 10, 11
Utley, Victor, 30

varves, 88, *89*
vein quartz, *77*, 77–78
Vitali, Gino, 74
volcanic activity, 26, 37, 45, 53–54, 113–16. *See also* igneous (volcanic) rocks; lava flows

Wadsworth, Daniel, 126, 130, 131
Wadsworth, Capt. Jeremiah, 67
Wampanoag people, 105, *106*
Washington, George, 101
water: lava flows as carriers of, 118–20; suburban/urban effects on supply, 2; as transportation system, 6, 120–23, *122–23*
weather: Connecticut's variations in, 27–28; deforestation's effect on, 2, 34; drought conditions, 2, 35, 40, 76; hurricanes, 6, 28–33; poetry on, 31, 36–37; political effects of, 36; tornadoes, 33. *See also* climate and weather
Webb, Joseph, 67
Weir, John Ferguson, 130
Werner, Abraham, 113
Western Highlands, 26
Weston meteor strike (1807), 161–65, 167
West Rock, *48*, 125–26, *127*, 128, 130
Wethersfield, Connecticut: European settlement of, 95, *99*, 99–100; meteor strike (1982), 6, 157–61, 167
wetlands, 118
Wheeler, Judge, 164
Wheeler, Nathan, 161
White, Henry, 138
White, James, 139
Whitney, Josiah, 69
Whittier, John, 108
Wilder, Thornton, 51
Willamette, Oregon, meteorite, 168
Williams, Eliphalet, 140
Williams, Samuel, 34, 134–35, 140
Windsor, Connecticut, 95
Winslow, Edward, 94
Winthrop, John, IV, 140
Winthrop, John, Jr., 57–61, *58*, 66, 75
Wisconsinan ice age, *41*, 45–49
witch hunts and Moodus noises, 135
Wolcott, Gen. Oliver, 70
Wright, Benjamin, 121

Yankee Power Plant, 150–51, 156
"year without a summer" (1816), 34
"year without a winter" (1826), 37–38
Younger John Winthrop, The (Black), 57–58